THE QUANTUM WAY

*Understanding the Science Behind
Happiness and Workplace Engagement*

Clive Hyland

Happiness and Humans Publishing

Copyright © 2021 Clive Hyland

All rights reserved

Thank you for your interest in this book and its material. If you would like to reproduce or transmit this material in any form or by any means, electronic, mechanical, photocopying, recording, or otherwise, for non-commercial purposes, please feel free to do so but do give credit to the author.

Cover design by: Joe Wedgwood

*To the newest members of the family who have arrived
since my last book, namely Claudia, Alex, and Zac*

CONTENTS

Title Page
Copyright
Dedication
Foreword by Matt Phelan
Author Preface
Chapter 1. Introduction — 1
Chapter 2. The Quantum Realm — 8
Chapter 3. Brain and Body Intelligence — 22
Chapter 4. The Human Experience — 40
Chapter 5. Self-Discovery — 57
Chapter 6. The Quantum Organisation — 70
Chapter 7. Happiness and Engagement Data — 103
Chapter 8. The Science of Belief — 115
Chapter 9. Transforming Society — 135
Chapter 10. What Can I Do Now? — 143
Bibliography — 149
Recommended Reading — 152
Acknowledgements — 155
About The Author — 157

FOREWORD BY MATT PHELAN

Hey everyone,

Welcome to *The Quantum Way*, book two in the Happiness and Humans series.

The Happiness and Humans series is an independent group of books written by independent authors connected by a shared vision that we call 'freedom to be human'. Freedom to be human is a way of working that focuses on helping individuals and organisations thrive together.

One of the barriers to achieving our vision is the old command and control structures that currently exist in many organisations. It is easy to say what is wrong with command and control, but it is much harder to suggest an alternative based on data, science, and evidence. I believe organisations have succeeded in spite of these archaic structures and not because of them. We need to find a new way.

We know from the data in our employee happiness platform across 90+ countries that many employees – and therefore their companies – are at breaking point and something needs to change.

This book is the sequel to my book *Freedom To Be Happy: The Business Case for Happiness*, but both books can be read independently and out of sequence. *Freedom To Be Happy* finished by outlining the quantum way, and since then we have been inundated by requests for more details on the why, what, and how of the quantum way. This book aims to answer these questions.

When Clive Hyland first introduced the quantum way to us his words were: "Here are some ideas: now you run with it!"

The information Clive outlines in this book blew my mind the first time I was introduced to it, but the simplicity of the ideas is what makes it so great. Words like quantum mechanics and neuroscience might sound scary (well, they did to me), but the best advice I can give you as you read this book is to remember that neuroscience simply helps us understand the way we work as human beings. Quantum physics helps us understand the universe we inhabit. Somewhere in between are happiness and engagement.

Later in Chapter 7, I will return to present the latest data and insight on how the quantum way works in the real world.

I will now hand you over to my coach of 12 years, colleague, and good friend Clive Hyland to explain the quantum way and the science behind happiness and engagement.

Over to you, Clive.

Matt Phelan is the co-founder of The Happiness Index, host of the Happiness and Humans podcast, and author of *Freedom To Be Happy: The Business Case for Happiness.*

AUTHOR PREFACE

I needed a nudge to write this book. I certainly felt I had something I wanted to share, but writing a book can be quite an isolating experience. Having done it twice before, I knew what I was letting myself in for. Then came nudge number one – Covid-19. Hardly the most pleasant reason, but as I was going to be stuck in lockdown for significant periods, the opportunity was timely. The second nudge, a far more enjoyable experience, was working with The Happiness Index. I have acted as their neuroscience and quantum advisor since their inception, and I am really impressed by the way they have fully embraced the principles embodied therein, made them their own, and ran with them with an irresistible sense of purpose. As such, this book is an acknowledgement to this spirit.

Matt Phelan, co-founder of The Happiness Index, recently published his book, *Freedom To Be Human: The Business Case for Happiness,* as the first in the Happiness and Humans series of anticipated publications dealing with themes which are central to our engagement in places of work. This book is the second in the series. Matt focused on research material linking productivity with happiness, and I will focus on the underpinning science, looking at what it takes to make us happy, what can get in our way, and the opportunity for personal growth and fulfilment that now sits before us.

For me, freedom and happiness are intertwined and will form central pillars of this book. Happiness can be described in different ways and typically involves a high energy state of enjoyment and excitement. This state is wonderful, but it is not sustainable. It involves a temporary energy spike that takes the body out of balance. We need to restore our natural equilibrium, and this can be described as peace or contentment, a more sustainable version of happiness. Both states are relevant in this book. Organisations need to be able to generate excitement at critical times, and they also need this to be underpinned by a sustainable contentment.

Freedom sits within this. The freedom to choose. Can we really be happy if we exercise little choice in our lives? Like happiness, freedom is a subjective experience. Each of us has our own special blend of needs and aspirations, and while freedom may look the same for us all at a distance, as we look closer, we can see the unique differences. One person's freedom may be another's prison. Sometimes we may face the evident challenges of being locked in situations we would not choose; other times our prisons are created within us, holding us back from where we would like to go. This book will not try to set out a happiness template that fits all sizes. We cannot be told how to be happy. Instead, we will examine the dynamics of the people we are and the choices we make throughout our life journeys. What causes us to make these choices? How free are we to decide? And what keys exist to open the prison doors?

I have always been a student of human behaviour, at various times studying sociology and psychology, before getting on to the 'harder' sciences over the last couple of decades. It runs deeper than just more formalised study. People fascinate me – how they 'tick' and perform. As a business leader, I always considered my greatest responsibility to be getting the best out of those who worked in my organisation. Early in my career I was exposed to the stimulating challenge of leading a team, and as I progressed into the corporate world, I had to learn to handle the scale of multinational businesses.

This is in stark contrast to my roots. I was brought up in the coal mining valleys of South Wales. What it may have lacked in terms of material benefits it more than made up for in terms of community spirit. The adversity of the mining environment meant that local people looked out for one other, not in any sentimental way, but with a robust honesty that let you know where you stood and kept you very grounded. I learned loyalty, friendship, and resilience, with my loving parents as the centrepieces in the unfolding drama. It was the environment in which I fine-tuned my social skills, especially the ability to be streetwise with the strong characters who shared those early years. This sense of connection and survival prepared me well for the career that awaited me. I look back on those friends with great fondness. They didn't all end up on the right side of the track, but they were good people. When I saw how tough some of their lives were, it left me with a life-long gratitude to my parents, who steered me so effectively to learn for myself, but with the knowledge that they were always there for me when needed. It also made me very wary of judging people without understanding their circumstances.

University was a real shock as it was the first time I had to deal with a middle-class dominated culture. I had to figure out the rules of the game quickly and soon found myself really enjoying the company of my new companions. This fascination with people continued both within my studies and, much more enjoyably, in the social activities that often followed. I suppose it was predictable that my corporate career started in Human Resources, where I worked in various roles for the first 10 years. Although enjoying many aspects of HR, I found the wider leadership urge in me irresistible and so moved across to direct business management roles, until I reached positions of COO and then CEO in the new millennium.

I look back on this career now with a mixture of pride and

pain. It wasn't easy to make the progress I achieved without making mistakes and meeting a number of opponents along the way. But I am convinced that the principles I held close were valid and honourable. I believe in people and their potential. Back then it was an intuitive thing based on my instincts and senses. More recently, I have been able to understand and articulate these beliefs in ways that others can more easily relate to. In this book I set out to share these insights with you. There is a whole new level of human potential waiting to be explored, and I trust the following pages will encourage you to embrace the opportunity.

Clive Hyland, March 2021

CHAPTER 1.
INTRODUCTION

This book is about reviewing our understanding of human experience and behaviour at the most fundamental level. We will go on a journey of discovery which will look at both human nature and the nature of the universe and will discover that ultimately the two are inseparable. We cannot understand one without the other. But we will not restrict ourselves to science alone. We will learn that intuition is a source of intelligence that should not be ignored. Readers will be encouraged to think well beyond where science can take us, to the world of subjective experiences and beliefs, where we shall learn that we are not alone.

Opinions or untested theories are not enough. We need to support the insights we share with credible and up-to-date science, but it must be science that is accessible and relatable. I have not written this book for scientists, although it must stand up to appropriate scientific enquiry. My target audience is much wider as I believe that this subject matter is far too important to be left solely in the hands of specialists. I have great respect for the scientists who are leading the way on this (indeed I would have nothing to write about without their guidance), but, understandably, their focus is not typically on educating Joe Public. So, I want to help fill the gap.

Before we get into some of the science, it may be useful to give a quick overview of the book. Once the initial scientific insights have been explained in Chapters 2 and 3, we will take a closer look at the human experience in Chapter 4, exploring how we perceive the world and the needs that drive us. In Chapter 5, we will push the boundaries further by examining what we mean by the 'self'. What does it mean to us? What is the scientific evidence to support such a phenomenon or is it simply an illusion? Chapter 6 brings us back into the more pragmatic reality of the organisation. What does a 'quantum organisation' look and feel like, and how does it differ from traditional organisations? In Chapter 7, we will change our focus to data. Matt Phelan of The Happiness Index will put in a cameo appearance for this chapter and will provide us with some fascinating insights into employee emotional states throughout the Covid-19 pandemic. In Chapter 8, we will allow ourselves some freedom to contemplate the wider perspectives of spirituality and beliefs and the role they play in our lives. Chapter 9 will look at the implications that quantum insights have for society as a whole and what it may take to grasp an opportunity for wholescale transformation. Lastly, in Chapter 10, we will return to the pragmatic and look at what we can do to take these insights forward and put them into action at both a personal and collective level.

The two main scientific disciplines we will refer to throughout the book will be quantum physics to understand the universe and neuroscience to understand ourselves. For many readers, these can be scary terms, but my ambition is to break down the – often self-imposed – barriers to wider understanding. Fortunately, the scientific community does all the grunt work for us, so our task is to understand the headlines and to put them within the context of our own lives.

Most people will have heard the term 'quantum', as in the general phrase 'quantum leap', but are unlikely to understand what it really means. While familiar to physicists, it rarely translates to everyday conversations. This is true of quantum physics in general, seen as a world for specialists only. And this is, in my opinion, valid at least in part. For anyone to grapple with the challenges of quantum mechanics, which is more concerned with the application of quantum methodologies, a serious maths or physics background is needed, which entirely disqualifies me. However, such a background is not necessary for the insights we will explore in this book. All you need is an open mind and a willingness to think differently about stuff you may well have hardly ever contemplated previously. In this case, ignorance is not bliss.

In a literal sense, 'quantum' means the smallest perceivable unit of any entity. But this is about much more than discovering the infinitely small: here we are exploring the fundamental building blocks of the universe, nature, and potentially of life itself. As we move our enquiry to the subatomic level and beyond, all that exists is energy. In discovering this quantum energy, we may have identified the source of everything that exists, including human beings. While neuroscience will help us to understand ourselves, these emerging quantum insights will inform us about the universe in which we exist. As we shall see in Chapter 2, ultimately, they are inseparable.

The label 'quantum physics' could either scare many people away as being too complicated or leave them cold as being perceived to be irrelevant to everyday experiences. Having waded my way through various books and many research papers on this subject, I can understand the feeling. I never set out to climb this particular mountain, and, indeed, I can only claim to be at base camp one; but, looking back, it seems inevitable that my journey would lead to this very special landscape.

Despite the challenges of unfamiliarity, I will argue that the insights of quantum theory are vitally important, translatable, and ultimately implementable in our everyday lives. I will present insights that challenge our old ways of thinking about what drives human behaviour. To endeavour to do this in an accessible way, I will strive to make the science elements as easy to understand as possible and to demonstrate their immediate everyday relevance. I am not a practising scientist; rather, I am a self-taught enthusiast who has spent the last 16 years working with such emerging insights, reading many relevant books (you can find my reading list at the end of the book), studying research papers, and talking with (and listening to) the scientists and medical professionals themselves. My role is one of translation from the sometimes very complex world of scientific methodology and language to that of practical reality and daily conversations.

Having stepped out of the corporate leadership world in 2005 to reconnect with my hunger for greater depths of human understanding, I found myself introduced to the world of neuroscience, apparently by chance – or was it? (I will return to the subject of synchronicity later in the book.) Having been gripped then and now by the fascination of brain and body intelligence, I also learned along the way that it is not just a case of what is happening within us, but equally important is what is going on around us.

The human intelligence system cannot be fully understood without examination of the energetic environment in which we live. We are comprised of energy and surrounded by it. We are part of a universe of energetic systems that not only impact us but actually are us. Our brains and bodies are constantly responding to this environment every second of the day, shaping our feelings and responses. While we might initially think in terms of the physical aspects of our environment, such as people and material objects, true understanding only comes as we equally address the energetic domain. It is indeed quantum physics which grapples with this dimension of the apparently unknown and invisible, phenomena previously dismissed as unprovable and matters of pure conjecture. Now we are finding that the material world itself cannot be entirely understood without these quantum enquiries.

So, I will introduce you to the emerging insights which have rocked my world. Neuroscience stands tall in my mind and heart as it unveils a new level of human understanding. At this point, the quantum universe may seem far removed from our understanding of the reality of everyday personal life, but such an assumption is wrong. The parallel insights of neuroscience and quantum physics have the potential to lift the veil of relative ignorance which has been blighting our view of universal and human nature for too long. We are so much more than that which we see and hitherto have learned to perceive. There is a whole new level of understanding to be revealed to those who have the courage and take the time to open their minds. Science can become a friend to us all, walking alongside us and helping us to see the world in deeper colours and feel life in richer textures.

This will appear strange to some who traditionally associate such life-enriching features with art much more than science, but I will challenge this. Indeed, challenging polarity of thinking and rigid black and white views will be a consistent theme in this book. The universe itself has no colour and is certainly not black and white; it is grey. Yet, we as human beings create colour in our minds through the magic of perception. So, working out what exists independently of us and what we create in our brains will be a productive line of enquiry as we look at the many barriers we construct, often unintentionally, to our own happiness and fulfilment.

The key for me is not to become slaves to one type of thinking. As this book will testify, I believe in the value of science, but it should not blind us to all the other forms of human intelligence that can be available to us. Convergent thinking offers us so much more than polarity does. The path of this book will therefore include a look at the profound convergence of thinking which is happening across previously polarised communities, stretching across science and art, spirituality, East and West, religion, and philosophy. All these sources of knowledge and inspiration deserve their place at the table of human knowledge.

The book will explore human behaviour in some depth at both a personal and collective level and will also take a particular look at the shape and operation of modern organisations. This will be integrated with my own understanding and experience of running organisations in order to construct a model which we can call the 'quantum organisation'. My belief is this proposition points the way for a new type of organisation which enables rather than denies the natural expression of human talent.

The book will also invite you to re-examine your thinking about yourself. What exactly is the self? What is the scientific evidence to support your current perspective? This will challenge you at the deepest personal level, but it is vital that we do so. We cannot simply look to our environment for change without examining the need to change ourselves. Our inner and outer experiences run hand in hand. While we can indeed re-examine the communities and organisations we have created, it would be futile to do so without understanding our own needs and responses at both a conscious and unconscious level. Consequently, self-discovery will form a major section of this book.

As we scan our understanding of ourselves, our communities, and our environment, we will discover through both intuitive and scientific enquiry some universal rules which apply at all levels of life. Despite appearances to the contrary, at the most fundamental level, the apparent differences between us disappear and truth itself merges into one.

The 'quantum way' is ultimately a vision of a better way of life that is accessible to us all but has been concealed by the limitations of our own thought processes. The combined sciences of neuroscience and quantum physics are revealing to us a whole new dimension of the interaction between humans and the universe. Central to this is the acknowledgement that we are, first and foremost, energetic beings. By learning to tap into this energetic potential, we can make a radical leap forward in human development, our own quantum leap forward. We can rebuild structures that support and enable our natural states and move away once and for all from those which constrain us, divide us, and fail to tap into our higher needs. This represents a completely new paradigm for human growth and connection. This opportunity is with us now, if only we dare to seize it.

CHAPTER 2. THE QUANTUM REALM

Background

While neuroscientific insights, a more familiar territory to me, will be woven throughout the book, I think it is worthwhile to get at least some understanding of the 'quantum' concept at this early stage, so we have some sense of what we are dealing with. This is key to understanding the new paradigm we are exploring in this book.

Let's go back to the definition offered earlier: quantum means the smallest perceivable unit of any entity. The atom was conceptualised in Greek philosophy and was identified in science in 1801 by John Dalton. The next two centuries saw further exploration of subatomic structures, including protons, electrons, and neutrons. At this level, such particles are composed of 99% energy. As this scientific enquiry continued into the 'infinitely small', unexpected insights were revealed. We discovered the quantum field, the essential fabric of the universe.

Whenever I have run client workshops on quantum insights, I have to take a deep breath: I know that explaining such insights will stretch both my communication skills and the thinking patterns of my audience to the limit. Invariably, I encounter at least some members of the audience who are very wary of a subject that appears to have nothing to do with them. On the contrary, it has everything to do with them, but I recognise that dealing with concepts as unfamiliar as these takes a big effort.

Quantum concepts ask us to set aside a lifetime of learning habits and start all over again.

The good news is that for the purpose of this book it is more a matter of grasping some of the key concepts rather than grappling with the complex formulae of quantum mechanics. Even the quantum physicists themselves say that anyone who says they understand the subject completely is demonstrating that they actually don't!

The shape of modern science, and classical physics, in particular, took a decisive step in the 17th century led by the work of Isaac Newton. His work on the behaviour of physical particles opened up an understanding of the material world at the most basic of levels of existence. This contributed to a golden age of scientific discovery and progress. The evidence of the importance of this era sits all around us, from engineering creativity, manufacturing processes, and modern technology. Our exploration of the physical world has paid off big time.

It wasn't until the early 20th century that some of the limitations of classical physics slowly started to emerge as scientific interest moved on to explore the next perceivable level of reality – the quantum or subatomic realm. So, let's look at one of the cornerstone experiments in quantum physics which started opening up a completely new understanding of the universe; this is the 'double-slit experiment'.

While widely referenced among physicists, it is rarely understood outside of this community, this being partly because of the difficulty of explaining it to those without a physics or mathematical background.

Imagine a screen with two vertical narrow slits in it, and behind this a second screen. The idea here is to examine the activity or behaviour of the smallest particle of light energy, the photon. Isolated photons are fired alternatingly through each slit in the first screen. Based on our normal perception of light we could reasonably expect to see two narrow vertical lines of light appearing on the back screen. However, the results are surprising. Instead of seeing two narrow lines, we will see two distinct interference patterns rather like the ripple effect caused by dropping simultaneous pebbles into a pond. Yet, this makes no sense as the photon was fired through the slot in isolation and therefore had no other photon to interfere with. It was as though the photon was splitting into two, each of which passed through their respective slits at the same time; but that had not happened.

To try to unravel this mystery, observation equipment was used (given that individual photons are far too small to be seen by the naked human eye) to explore what was happening as the photons passed through the slits. Then something weird happened: at the moment the observation device was switched on, the photon's behaviour changed to the original expectation, that is the clearly defined and separate vertical slits appeared on the back screen; the interference pattern had disappeared. There was no logical explanation for this, but the evidence was clear: the act of observation directly influenced the perceived behaviour of the photons. When observed, the photons behaved like particles; when they were not individually observed, they behaved as a light wave. Quantum physicists refer to this materialisation of the particle as the collapse of the energetic wave function.

This was groundbreaking because until then the Newtonian classical view was that particle behaviour was entirely deterministic: that is predictable, measurable, and linear. This experiment showed that there was a major flaw in this argument. It is a little like believing we understand somebody by looking at a photograph of them. We may know where they were and what they looked like at one point in time, but that hardly gives us the information we need to truly understand the person's behavioural patterns.

Something was happening which the science of the day could not explain. Einstein referred to quantum phenomena as "weird stuff at a distance". What emerged was the understanding that particle behaviour can only be understood as part of the wider energy system to which the particle belongs. In terms of light energy, the photon's behaviour is always part of the wider light wave and cannot be separate from it. No particle exists in isolation. Everything we perceive consists of both energy and matter. Nothing is quite what it seems.

Roll forward through the emergence of quantum physics methodologies throughout the 20th century and we currently see a very different view of the universe. At the most fundamental level of existence, everything is energy – more specifically, quantum energy. Matter itself is a materialisation of this energy, which comes about through the act of human consciousness. Energy and matter are two inseparable sides of reality distinguished only by human perception. Separation of energy and matter sits within us, not within the universe itself.

The universe itself as we perceive it does not exist independently of us. We create it in our own minds through the act of human consciousness. If we remove the role of conscious observation, the universe is an infinite realm of interconnected energetic systems that exist as a vast ocean of potential. The act of human conscious observation enables these energetic phenomena to materialise in our minds as physical entities which we can experience through our primary senses of sight, sound, smell, taste, and touch. But these entities only exist in this form for us. An entirely alien species from another planet may perceive and experience the universe in a completely different way.

Such discoveries inevitably undermined the once prevalent deterministic view of the universe. Science was acknowledging the limits to its own methodological enquiry. We now see the universe in a more probabilistic way, where we can predict certain physical outcomes with confidence, but we cannot be entirely clear about how they happen. This is illustrated at the quantum level, where electrons appear and disappear within their atomic orbit, apparently at random. Quantum reactions happen at a speed significantly quicker than the speed of light. We have no provable means of working out how this happens.

In parallel, the relationship between time and space has been redefined (Einstein's theory of relativity) and are no longer seen as independent phenomena: time can be 'bent' in space and our experience of time would change dramatically if we were able to travel near the speed of light. Quantum physics tells us that if we were to travel to a distant planet at around 99% of the speed of light and it took us 10 years to return, at the point of return, the people on Earth will have aged 70 years! Our apparently rock-solid assumptions about time and space are only a function of the limitations of our powers of perception. These rules do not apply beyond the realms of our immediate local, physical environment; they cannot be used as a means of understanding the universe. The universe is infinite, a concept that our minds struggle to process. Philosophers have grappled with these challenges over many centuries, in terms of both the cosmos (the large infinity) and the quantum dimension (the small infinity). We cannot define or discover where the universe begins or ends, what is the largest or smallest, what is the fastest or slowest. Such limitations exist only in our minds.

Chances are at this stage you are reaching for the ice bucket to cool your brain, but please bear with me as I try to anchor some of the principles that will have significance throughout the remainder of this book. They have much more practical significance than may at this stage seem apparent. In doing so, I will refer to a number of what I will call 'quantum features'. These simply represent my take on the stuff we all need to understand about the universe and the way it works. We can leave space travel to the scientists! This is not a publicly acknowledged list to which all quantum physicists subscribe; I doubt such a list exists. Nevertheless, each of them is a cornerstone principle within the world of quantum study and has relevance to all of us.

Let's look at them firstly within the context of the universe, and then we can start to bring in some everyday relevance.

Key features of the quantum universe

Everything is energy

Physical matter accounts for less than 1% of the known universe. As we drill down further and further into the quantum nature of matter we are left only with energy. Subatomic particles such as protons, electrons, and neutrons are energetic units that materialise only through observation. The universe itself is an infinite field of energy and this energy has unlimited potential. Energy exists in a variety of forms, including gravity, heat, chemical, radiation, and light. Of particular interest to us is electromagnetic energy and its role in the human experience.

Ultimately, therefore, everything is energy, and this includes humans! Yet, in the Western world, we have paid so little attention to the energetic essence of our existence. This has blinded us to many insights about ourselves. This book will start the process of unravelling these hitherto unknowns.

All energy is connected at the quantum level

Nothing and nobody exist separately from the quantum universe. The implications of this are enormous. Here we have a scientifically endorsed view that we, as human beings, are connected energetically. Nobody exists in a vacuum, and we follow the same behavioural principles as quantum phenomena in that our lives are constantly influenced energetically by what is going on around us. We cannot define or experience ourselves in isolation from others. Furthermore, this connection extends to all aspects of nature, including animals and plants.

The universe is emergent

We now see the universe as emergent, that is unfolding. A deterministic view of the universe has therefore largely given way to a probabilistic one. The concepts of emergence and probability go hand in hand. We can be confident that we can predict some outcomes (think about the precision of space travel), but we cannot be entirely sure how it works. While we have some understanding of matter, our understanding of the remaining 99% (energy) is very limited.

The universe operates at the edge of chaos and has not yet exposed its secrets to rational enquiry alone. Far too much information is inaccessible and beyond our experience. So, any conclusions we draw about the universe have to be probabilistic; that is, we can't be completely sure. Importantly, this 'uncertainty principle' creates room for the acknowledgement of other forms of legitimate enquiry, including intuition and spirituality.

The universe is self-organising

The energy of the universe is self-organised into an infinite number of systems which are linked through energetic resonance. These systems are sometimes referred to as 'fields of meaning', where they are connected through some central bonding force. When the Big Bang created the universe some 14 billion years ago matter was created for the first time. The universe continues to expand and is constantly making new energetic connections and forming new paths to maintaining its momentum. We will examine how this self-organisation can work for human communities.

Energy operates in flow
Energy flows. It can be neither created nor destroyed; it can, however, be channelled or transformed. It cannot be boxed up. It is the life force that we channel through our daily activity and experiences. Understanding flow has significant implications for human expression, fulfilment, and performance.

All of these 'features of the universe' apply in equal measure to us. We are literally children of the universe. The chemical elements that we are comprised of were formed by the fusion of the stars in the infancy of the universe. We do not occupy the universe, implying some essential independence from it; we are part of it, and it is part of us. The universe runs through our brain and bodies as an ever-present and active force.

Human impact
Let us now take some first steps to anchoring these unfamiliar concepts into the human experience.

Everything is energy
We experience energy as part of our everyday lives primarily through our senses of sight, hearing, touch, taste, and smell. Yet, we remain preoccupied with a psychological emphasis on mind. The implicit assumption here is that we are mind-centred beings. While there is undoubtedly value in this outlook, it also has important limitations. We are first and foremost energetic beings. Energy is our essence and the source of who we are.

The role of the mind is to help us make sense of our lives and to make conscious choices about how we move forward. Yet, 95% of our behaviour is triggered at the subconscious level and precedes any involvement of the conscious mind. These are energetic triggers which set the pattern of our responses, both consciously and subconsciously. Our whole interpretation of the world is far more subconsciously driven than we think, a subject we will examine in more depth when we explore human perception in Chapter 4. The conscious mind is like the jockey on the horse: it has a vital role in harnessing and guiding the energy of its companion, but it is never in complete control.

All energy is connected at the quantum level

My guess is that all of us at some time have experienced that special sensation we feel when we connect with nature, when we feel the touch of the wind against our skin or the depth of the ocean talking to our soul. There are many spiritual views on this, but here we can offer a quantum insight. Everything around us is connected at the quantum level, and when the barriers of concrete and clay are stripped away and we allow ourselves to be in the moment, we can sense this deeper sensation.

The same applies to connection with animals, especially mammals. When we are calm, present, and look into the eyes of a horse that has learned to trust us, it is like a moment of love, a deep mutual acknowledgement of each other's value and role in the universe. Over the years I have been a dog lover and I will always cherish the depth of unquestioning bond that can be shared in special moments when we just know that we are there for each other. This is a quantum connection, where our personal energies resonate at the same frequency, and we tune into each other's deeper selves.

Have you ever wondered at the marvel of birds flying in synchronisation, such as starlings gathering in their thousands and flying in instantaneous rhythm as they dash and dart across the sky? How do they do this? How can they respond to each other with such precision and at such speed? We now understand that this is achieved through quantum signalling.

We have evidence that plants respond to human love. In properly controlled experiments, plants that were consistently exposed to positive human intentions thrived better than those that were not. We also see that trees operate in communities: when one tree is sick, the other trees will reduce their intake of nutrients from the soil to allow the ailing tree the chance to recover. How do they sense this? If you place a plant in a vacuum, it will die. If you place a mammal in a vacuum, it will suffer the same fate. Yet, if you place them in the vacuum together, they will both survive. This is quantum connection.

Regrettably, our approach of recent centuries has taken us down the path of individualism and egocentrism, where we have been encouraged to see ourselves as separate from others. The self is seen as a constant essence that navigates its way through our life experience. In fact, the opposite is more likely to be closer to the truth: our experience of life defines our essence, and this essence cannot be developed in isolation from others.

We have paid a heavy price for this sense of separation. We have seen our environments as places we occupy and nature as extrinsic to our being. All too frequently our disconnection with others leads to further withdrawal and depression. It is no coincidence that a deep sense of isolation is usually the last step before suicide.

The universe is emergent
I explained the concepts of emergence and probability earlier: there is always an element of uncertainty in our understanding of the universe. The universe is an infinite field of unlimited energetic potential. Its complexity stretches well beyond our comprehension; therefore, we cannot predict how this energy will manifest itself.

The difficulty for us mere mortals is that we, especially in the

West, have been educated almost exclusively in linear thought: the world of rules, methods, and structures. But none of the universe's secrets can be explained by classical physics alone, where we use rational processing as the dominant force in our thinking. As we shall see in more depth later, rational processing engages primarily the left side of the human brain and is largely deductive, but to get our heads around this stuff we need to engage the whole of the brain. We need to exercise our instincts and intuition, and we need to activate the right side of our brains to trigger creativity and inductive thinking, combined with the visual power of imagination.

Newtonian physics was beautifully equipped to deal with the external physical world but completely set aside the world of our inner experience (what scientists call 'qualia'). We now understand that this is like seeking only to explore the bright side of the moon, solely that which we can see in the physical world. That which we cannot see is part of the same system and equally vital to our understanding.

So, a quantum approach has to be holistic; we cannot understand the whole by a reductionist analysis of each of the parts. It is the combination and integration of the parts at an energetic level which makes the whole what it is. We cannot understand table salt (sodium chloride) by breaking it down into its constituent parts. Sodium is an explosive in water and chloride is a poison. I don't fancy the idea of putting an explosive poison on my food! The classic metaphor here is the Frankenstein myth: no human being (or man-made equivalent) can be understood by its body parts alone. We need to access the energetic life force that is its essence.

The universe is self-organising

I have explained that the universe evolves through a process of ongoing connection and reconnection of energetic fields. We have witnessed this here on Earth in the form of evolution. There is an undeniable evolutionary force that has shaped our world slowly but surely since our planet was formed around four billion years ago. This 'life force' cannot be turned back; it keeps moving forward, relentlessly and patiently. We may not perceive an endgame, but evidence of the journey is a matter of observable fact. If one potential way of evolutionary progress is blocked, it will find another. If the human species were wiped out in some apocalypse, evolution would recover and try again, just like it did when the dinosaurs were destroyed, and small reptiles and mammals took their place. Within this context, we are not talking about a precise plan, but rather an indomitable force that will not be denied.

The term 'exaptation' is familiar to evolutionary biologists. It is when a particular competence is repurposed to do something else. Here are two examples of such evolutionary ingenuity which have benefited humans throughout the ages. The delicate bones of the human inner ear can be traced right back to reptiles, not to their ears, as you may expect, but to their jaws. Early lizards rested their jaw bones on the earth to be able to feel the vibrations of other animals in the vicinity. These bones were exapted to use their sensory potential in the human inner ear. In parallel, human sight can be traced originally back to plants. The process of photosynthesis involves plant cells converting light energy into nutrients essential for their survival. Down the line, this ability to process light came in very handy when it was repurposed to processing human sight!

My point here is that evolution itself is a manifestation of the universe and its power comes from the same force. It is incredibly resourceful, connected, purposeful and resilient, just like the source from which it came.

Energy operates in flow

The principle of flow has significant importance for our experience of human life and the feeling of happiness. As explained earlier, flow is a natural state of energetic momentum. It cannot be boxed in; any attempt to do so will simply build up energy until it finds a way to release itself from the confines of its entrapment, either through gentle escape or violent explosion. And so, it is across nature: flow exists when nature is at ease with itself. The energetic force of flowing water cannot be stopped. We can build dams to moderate it, but we have to work with it, not against it.

Flow in human beings is vital to our enjoyment of life and our ability to perform. Artistic expression is a clear example of flow, when we allow our inner world to be shared with others, where the self-constructed barriers between inner and outer experiences fall away. Art is fluid and associative, just like the natural energy force from whence it came.

We will look at this in more depth in later chapters when we will explore the ways we build up resistant energies in our own lives leading to stress and illness. We will contrast this with the natural state of human flow, such as that which is labelled the 'zone' in sporting communities, and we will discover the enormous leap of untapped human potential available to us both individually and collectively.

Most of us recognise how important a sense of purpose is in our lives (and in our jobs!). The human experience of purpose is a natural state of flow where our energy is aligned with our intention and our direction of travel, where we become entirely engaged with the cause that is attracting us, our 'field of meaning'. The universe is designed to flow, yet we have unknowingly developed human communities and organisations that are resistant to and disconnected with these natural states. It's time to think again!

Hopefully, this chapter has given you some sense of quantum concepts and how they relate to everyday life. In the next chapter, we will turn to neuroscience as the parallel discovery track that we shall be using to explore this emerging and radical view of human life, and what it may take to get the best out of it.

CHAPTER 3. BRAIN AND BODY INTELLIGENCE

Background

Although an equally complex subject, neuroscience is nevertheless less conceptual, more grounded in the tangible, and more relatable. References to neuroscientific insights will recur throughout the book as the parallel science track we are interested in here; but, for now, let us just lay a platform of understanding by looking at its history.

Although a newly emerging science, neuroscience has been with us throughout the 20th century. At that time, much of the research was focused on animals as the technology did not exist to give us access to live human brains, although dissecting dead brains was useful for building an understanding of brain anatomy. This changed in the 1990s when the development of functional MRI scans and other imaging technology allowed us to see live human brains in operation.

Psychology flourished in the last century but was equally hampered by live access to the operating brain. Consequently, psychology relies on a theoretical model where we build the 'theory of mind' without being able to define exactly what it is. It therefore relies heavily on behaviour observation and analysis. While this brought about major advances in our understanding of human behaviour, neuroscience now offers the next level of discovery. We can certainly demonstrate significant progress in our understanding of the machinery of the brain, although areas such as 'mind' remain more elusive.

This is where the combination of neuroscience and quantum physics starts to unveil the mystery. For now, we can think of the brain as the physical organ which sits within our skull with its mass of brain cells. The mind, on the other hand, sits very much in the quantum energetic realm. It has no physical properties, but we intuitively know that it is real and has a major role in directing our lives. This is a beautiful example of where the quantum and physical dimensions meet. The brain cannot be accurately understood without reference to the energetic phenomenon which is the mind. Equally the mind cannot be understood without reference to the brain. They are physical and energetic counterparts that have meaning only when they can be understood as one system. The one cannot exist without the other. This is a very important insight which is central to the new emergent scientific understanding of our perceived reality.

The brain

At this point, let us give ourselves a basic understanding of the brain and some of the key principles. I have covered this subject area in my last book, *The Neuro Edge*, but a quick summary is useful here.

The brain weighs about 2 kilos and consists of between 90 billion to 100 billion neurons (brain nerve cells). When we are born these neurons already exist, but, crucially, they are not yet connected. If our brains were fully connected, our responses to life would lack flexibility. At its outset, the brain is an organ of enormous potential. The neurons are in place and ready to respond to the environment into which it is born. The process of connecting the neurons starts in the womb which is the first point at which we start to receive 'life data'. Mother's heartbeat will be one of the earliest experiences for the foetus and will already start to prepare the baby for the world to come. Our brain is like a computer programme that is ready to go, and then comes the life data needed to activate our capabilities.

At the point of birth there is an explosion of connectivity as the baby starts the process of collecting information from its external environment. The neurons connect through chemical messengers, known as neurotransmitters. These junction points are called synapses, which enable multiple and varied connections with other neurons in the same proximity, turning on and off as and when needed. Synapses enable the brain to maintain its flexibility throughout life. While we will have a broad template for the functioning of our brain at birth defined by our DNA, it is the process of interconnection between the neurons that customises our own brain to the environment which will condition our own unique experiences.

This process will continue at an incredible speed in the early years at about half a million new connections every second. Rapid interconnection will continue until about 12 years of age when adolescence kicks in and the brain starts to prune itself of the connections it no longer needs. This re-engineering continues through our teenage years and the brain reaches biological maturity at around 24 years of age. Maturity here means that the basic biological templates are fully functioning; it does not mean that the brain is no longer flexible. The generation of new brain cells ('neurogenesis') and constant interconnection carries on throughout our lives and is called brain plasticity. This is crucial to ongoing health and learning.

The brain is a predictive machine. It constantly anticipates the future, adjusting as we go along. A little insight on this happens when the unexpected happens. For example, when you approach an escalator your mind is already ahead of your actions. It receives and assimilates sensory information through peripheral vision and interprets that you are about to step on to a moving escalator; accordingly, it adjusts the brain's sensory balancing system to get you ready to move in synchronisation with it. If the escalator isn't working and you haven't noticed, the first steps you take will be disorientating because the prediction was wrong and the time for adjustment was very limited.

Of course, even a summary description of the brain could consume many books, so we need to target our interest here. Many people will have a broad awareness of the difference between the left and right side of the brain, with the left being primarily concerned with logic and the right with emotion. This is true but we need to deepen this understanding, and this requires us to look at the hierarchy of the brain and the way that different capabilities have evolved throughout our history. An explanation of these core systems will help us to make the link between the inner dynamics of the brain and the human behavioural patterns we observe in everyday life (see Figure 1).

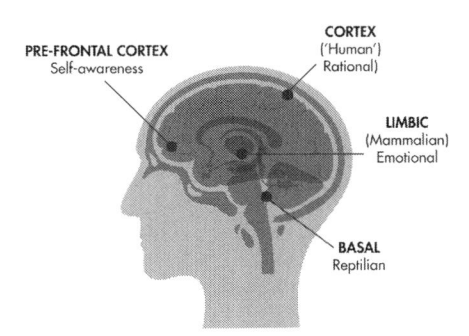

Figure 1: Brain Evolutionary Stages

The brain's behavioural systems

Human behaviour is the result of a very sophisticated chain of activities arising from external triggers and the responses we have learned, largely subconsciously, to get us through life. These responses are triggered right across the brain, so any attempt to localise behaviour to particular regions has to bear this note of caution in mind. Ultimately, the brain functions through a multidimensional matrix of physical and energetic capacities: nevertheless, there is a great deal of practical value in understanding the brain regions which form the source of the behavioural styles I will describe.

We refer to thoughts, feelings, and instincts as part of our everyday conversations, but a better understanding of how they happen in the brain will help us trigger new insights which will be extremely useful as we move forward. We cannot 'see' (through imaging technology) different phenomena in the brain which we can separately identify as thoughts, emotions, and instincts (essentially, they are all electrical impulses racing around our neural circuitry), but we can see which areas of the brain are activated in processing each of these impulses. This helps us build the model explained in this chapter. Furthermore, while these regions can be legitimately identified as the source of particular behavioural patterns, the brain does much more than direct our behaviour. We are focusing here on the aspects that are easily translatable into our everyday lives.

The model we shall use is a direct representation of the evolutionary hierarchy of the brain. You will have heard of general references to the left and right sides of the brain – and we shall take a further look at this later – but it is more revealing at this stage to look at this hierarchical model:

- The basal brain system and **instinctive** behaviour
- The limbic system and **emotional** behaviour
- The neocortex and **rational** behaviour
- The prefrontal cortex and **reflective** behaviour

I have found this model of behavioural styles to be highly engaging and productive for the clients I have worked with over the last 15 years. It provides a very practical and accessible entry point for starting to get to grips with the dynamics of our brains in a way that is relatable to everyday life. However, before we go into each of the systems and styles, there are few key things to note and explain.

Firstly, there is no wish here to box people into only one or two styles. Our brains are far too flexible and sophisticated for that. Our styles will respond to the environment we find ourselves in, so there is always a situational element. Our behaviour will inevitably be a blend of these styles, and we can switch between them in seconds. Nevertheless, an overall pattern is likely to emerge where, for instance, one style is used across all environments more commonly than others. It can be very insightful to explore this dynamic in understanding ourselves and those we interact with.

Secondly, because we show tendencies towards some styles more than others, it does not mean that we are unable to use the lesser styles. This is a preference model, not a competence assessment. These preferences have been subconsciously learned through life. As our brains have continued their adaptation to our environments, they have learned to activate certain response regions in the brain more than others. It is rarely a conscious choice. And just because we don't use a particular style often does not mean we are unable to do so; it may be a case of being able to use it if we really have to, it just isn't our preference.

Thirdly, there are no good or bad styles; they are simply what we have learned. As we shall see, each style has its strengths and its weaknesses. The challenge is to know how we get the best out of our own unique personal blend and how we can most effectively interact with the behavioural blends of others.

The basal system: the instinctive brain
The oldest part of the brain in evolutionary terms is the basal system, which sits at the base of the brain just above the brain-

stem and includes the pons and cerebellum. It is responsible for running many of the autonomous (automatic) systems of the brain and body, including breathing, balance, blood circulation, and digestion. It functions like our automatic pilot and is not dependent on conscious input to operate. The cerebellum is the most intensively interconnected part of the brain and incredibly sophisticated in its activity.

Our instinctive responses are very fast and potentially explosive. These reactions are driven by our instincts to survive and thrive. There is no point in having a survival response that is slow and weak. If it needs to engage to address an immediate threat, the oxygenated blood supply of the brain is diverted to focus on the motorneuron areas of the brain: in the moment, it is all about acting to get away from or take out the threat. When the basal system is highly activated by a significant threat it takes control, and other capacities – such as perspective, objectivity, and creativity – are temporarily discarded in the moment as being irrelevant.

The instinctive part of the brain is often referred to as the 'reptilian brain'. A visual reference can be helpful here to help us make the link with behaviour: think crocodile! Apart from being an example of a long-surviving reptilian species, the crocodile has easily recognisable behaviour patterns. For this creature, life is simple: it revolves around killing and eating, sleeping and resting, and reproduction. Yet, it is magnificently honed to survive in its natural environment. They are not social animals, collaborating with other crocodiles only to complete the occasional kill and tear flesh. They are not trainable and do not negotiate. As the song goes, "Don't you ever smile at a crocodile!" They are either explosive in executing their kills or completely still, absorbing energy from the sun.

So, why is the croc of so much interest here? Because it is an excellent illustration of the way that people with a strong instinctive style behave. Such a style is quick and decisive when engaged. It adopts a position quickly and cares little for how others are impacted by that position. It doesn't sit on the fence. It is certain in its actions and is either 'all in' or 'all out'. Its impact is like marmite: we either love it or hate it. It hates fuss and prevarication. It gets bored easily if it isn't quickly engaged. If you are dealing with such an instinctive style, get to the point, don't take time building careful evidence or rational argument, and forget the fluffy emotional stuff! A person with a strong instinctive style will make their mind up quickly anyway, so any time wasted is only likely to antagonise them. They portray confidence and certainty and can be seen as powerful. On the other hand, they stop listening quickly: why would they waste energy listening further when they already have the answer? Hence, a very thin line can exist between confidence and arrogance.

People with a strong instinctive style get noticed because they typically do not typically hold back. They are comfortable with taking the lead. The instinctive space is a selfish space, and this is said without judgement. First and foremost, our instincts are fine-tuned to look after us and the survival of our genes. It is not a style which naturally embraces emotional sensitivity and team play.

The limbic system: the emotional brain

The next distinctive evolutionary stage of human brain development was the emotional region, situated in the middle of the brain. Our emotional capacity, like all evolutionary developments, is a response to environmental demands. Evolution had worked out that it would be advantageous to work in groups: hence the emergence of mammals whose survival against reptile predators came to depend on their ability to cooperate. The purpose of our emotions is to bond us into a group or family entity. This 'mammalian brain' developed highly sophisticated human technology to enable relationship bonding and collaboration at a much more superior level. We consequently saw the emergence of social groupings and hierarchies as mammals instinctively understood the need to keep close to their groupings. Think gorillas, monkeys, humans, wolves, and so on. Social roles emerged, such as the alpha male, as mammals developed the structures best suited to survival.

The emotional brain is incredibly complex. It includes the ability to communicate through body language and posturing as well as facial recognition. We subconsciously transmit thousands of facial micro-signals every second of our interaction with others. This subliminal communication is received via our resonance circuitry sitting at the front of the limbic system, where we translate these signals into something meaningful to us. This forms the basis of empathy and trust or distrust and fear. This system of the brain seeks connection with others. It is typically tactile and demonstrative.

It is also the creative centre of the brain. Although other parts of the brain will collaborate to develop an idea, the emotional system works in 'flow'. It is an energetic force which does not like being boxed in. This energetic field gives it a capability to detect thought and feeling patterns across the brain and to assemble them into a meaningful jigsaw. We can see this in our daily experiences: art and emotion are inextricably intertwined. Deep emotional experience is usually the critical spur to artistic expression. Art itself is a subjective expression guided by our inner world.

People with an emotional style can be easily spotted. They are activated by their energy in the moment. They are naturally expressive, wearing their hearts on their sleeves. They are animated and can be enthusiastic one moment then feeling down the next. This is the expression of energetic flow, either up or down. Apart from being creative, their strengths are typically in team play, building relationships and energising those around them. On a bad day, they can become high maintenance and easily distracted.

How do we best engage with this style? The processing style of the emotional brain is associative, as in the creative description given above. This means that such a style engages best with storytelling. Stories are typically associative, flowing sequences captured in a narrative, a visual image, or a song. It is less about getting to the point; the emotional style wants to enjoy the experience. It's more about the journey than the result. This behavioural style values the connection of a relationship and wants to feel acknowledged as a person. They are energetically tuned in to those around them and will be very susceptible to the moods of others.

The neocortex: the rational brain

So far, we have briefly tracked the evolutionary development of the reptilian and mammalian brains. Next comes humanity. What separates us most of all from other species? It is our ability to think. While we take our thinking capacity for granted, it is the source of our dominance of our global environment. The 'thinking brain' sits primarily in the neocortex, which is the grey matter situated at the top and sides of our brains. While other mammalian species have a cortex, they have nothing like the sophistication available to human beings. Essentially, other species do not think, at least not in the way we can.

How did this evolutionary breakthrough occur? We really should sit up and take notice at this point, shouldn't we? Actually, the explanation is quite low-key. This incredible breakthrough in our evolutionary development happened when we learned to slow down! Instead of responding at the breakneck speed of our instincts and emotions, we learned to slow down our responses. This apparently minor increment in processing time gave us a whole new capacity to consider wider information before we responded. It takes 80 milliseconds to register an emotion in the brain and 250 milliseconds to register a thought. The difference is significant in the world of brain speeds. More processing time allowed wider information to be considered, and, step by evolutionary step, we learned to exercise choice. While instantaneous reactions are vital in life threatening situations, considered responses allow us to get ahead of the game. It is ironic that learning to slow down in the moment has helped us to speed up the progress of the human race.

The neocortex is the region where we lay down the neural circuitry that captures the learned rules for conducting our lives. Whereas our instinctive responses are largely influenced by the expression of our genes, the rules in the neocortex are learned both consciously and subconsciously. This neural wiring acts like a digital network directing electrical impulses through a series of decision gates (synapses) to trigger a response.

In this model, I refer to this as the rational brain. This is where our capacity for logical thought processing is centred. It seeks out order and clarity. It is comfortable with mental constructs like structure, processes, and systems. It lays down the rules in its neural circuits and follows them steadfastly whenever it can. The processing style here is primarily deductive, that is the analysis of information to come to a logical conclusion.

People with a strong rational style will present themselves in an orderly fashion and be considered in their behaviour. They prefer to think things through before sharing them. They like the structure of a methodological approach and seek out evidence and facts. They are reliable and diligent and like clear rules which they are happy to stick to. On the other hand, they can be perceived by others to be as emotionally disconnected or slow to get to the point. Classic examples of rationally driven professions are plentiful, including accountants, lawyers, researchers, and scientists.

The prefrontal cortex: the reflective brain
The fourth evolutionary stage was the development of the prefrontal cortex situated at the front of the human cortex. Although part of the wider cortex, this is the youngest part of the brain in evolutionary terms. What makes it distinguishable from the rest of the cortex is the direct neural access it has to all parts of the brain. In this way, it is able to assemble instinctive, emotional, and rational data to construct a complete view of our own inner experiences. In this sense, it is the source of our self-awareness.

The prefrontal cortex can be considered as the window to the rest of the brain in terms of normal social interaction. This is the region of consciousness, whereas the rest of the brain primarily operates at the subconscious level. It guides our voluntary actions and collaborates with all other regions of the brain to undertake complex human activity. It combines with our instincts to develop our intuitive sense; it works with the emotional brain to make sense of our experiences and to develop creative ideas, it brings innovation to the rationale of the neocortex to give us deeper insight. The list is endless.

While still broadly a thinking style, it is quite distinct from rational processing. Rational thought is deductive and follows rules already established in the brain. The thinking process of the prefrontal cortex has more freedom and is reflective and inductive in that it takes an idea and builds on it like the assembly of a mental jigsaw. This is imagination at play. From here we view the bigger picture, the wider horizon and the broader context.

This capacity needs a little more time to bear fruit. It takes about 420 milliseconds for a thought to register at the conscious level. We can recognise this ourselves in our own lives. To be reflective we have to give ourselves some emotional space and mentally temporarily withdraw from the immediate situation. We become the director surveying the scene rather than the actor immersed within it.

Above all, it is the place in the brain where we try to make sense of our lives, where we seek meaning and purpose, and where we search for and set a direction for our lives. It is the source of our motivation and our striving for personal growth and self-fulfilment.

Within the context of this model, this is called the 'reflective' style. People with a strong reflective style like to take in the world around them. They look past the obvious and look for something more. They see the patterns that others may not see. They are innovative and intuitive. They have the capability to be visionary in outlook. As such, they can be insightful, strategic thinkers, viewing the world from the helicopter rather than the trenches. On the other hand, they can be viewed by others as academic, self-indulgent, or lacking in an understanding of practical reality. It is hard to have your feet on the ground when you are flying a helicopter!

To engage effectively with such a style, it needs to be given the space to explore and innovate. It does not respond well to instruction or rigid methodology. Ideas may range from the brilliant to the nonsensical, from the immediate breakthrough to the longer-term conceptual navigation of the strategic landscape. These are people who can be time-expensive, but who also have the potential capability to spot the next big paradigm shift.

All the styles!
The point of this model is to give us a framework for heightening our awareness of human behavioural dynamics that links directly to an understanding of the brain. We need a means of understanding what is going on between us and inside of us. Separation of the styles has to be a conceptual process. The human brain will not limit itself to predictable responses and we will always use a blend of styles; however, I know from the last 15 years of my personal experience that understanding these components of behavioural patterns brings important insights that can allow us a fresh opportunity to move forward more effectively.

The magic of the brain lies in its capacity to integrate these styles and match them to the situations we face. Few of us would say that we have perfected this process. We regularly encounter situations where we become upset by the trivial or disappointed by our own reactions, where we are held back by our fears or fail to perform to our best. So, the journey into the alchemy of the brain continues.

The wider human intelligence
Our bodies are often assumed to have simply a physical role while all intelligence is confined to our brains, but this is totally wrong. We must remember that, in evolutionary terms, we had bodies before we had brains. As we track our evolution back to well before the reptiles, we discover simple single-cell organisms that had no brains, but they did have intelligence. Each cell had its own sensing mechanisms which enabled it to sense movements in its environment. Our brains emerged as our ambition grew to move to wider environments. The brain is essentially an intelligence hub developed to handle and integrate the increasingly complex information needed to survive what eventually became the human journey. Our brains are part of a

network of distributed intelligence centres which exist across our bodies. It relies on the exchange of information with these centres to be able to function.

The heart
Let's start with the human heart. The heart is so much more than a sophisticated pump circulating blood around the body; it plays a crucial role in human intelligence. The heart is formed in the foetus in the womb before the brain. It has its own level of independence and does not rely on the brain for instruction. In fact, the heart sends more signals along the vagus nerve to the brain than the brain does to the heart. It directly impacts the performance of the brain both at a neural and energetic level. It has its own intelligence system, referred to by neurologists as the 'cardiac brain'. This consists of about 10 million nerve cells, much less in number than the brain but significant in acting as something of a neural motorway system providing a direct connection between the heart and brain. Additionally, the heart plays an important role at the electromagnetic level: electromagnetic waves are transmitted across the brain and body and help synchronisation of intelligence activity.

In particular, the heart is directly connected to the limbic system of the brain and is highly influential in our emotional experiences. While the brain is the centre of our integrated and higher intelligence, the heart is the centre of our energetic connectivity to the external world. It is a key part of our sensing capability which impacts not only feelings in the moment but also areas such as trust judgements; this is where we come across the phenomenon of magnetic resonance.

In Chapter 2, we discussed the parallel existence of both physical matter and energetic forces. The heart is an example of an organ which functions on both levels. We can readily understand its very tangible role in the physiology of the body, but it also has a role in connecting us to others through energy. Electromagnetic waves are constantly exchanged between all species of life. These waves carry encoded information at different frequency levels, like tuning a radio into a broadcasting station. The connection is created when transmitting waves are received at a point of resonance, where they are able to exchange information. The heart does exactly this and is constantly tuning us into the energetic waves being transmitted around us by other living species, particularly by other human beings. This contributes to our emotional sensing capability to understand mood and the atmosphere that can exist between people. This happens instantly without any conscious direction from the brain.

The gut
The gut is also a major player in human intelligence. It has its own neural hub comprised of about 100 million nerve cells and is directly connected to the basal region of the brain. We often refer to experiences like 'gut feeling' to explain a hunch, an instantaneous intuitive sense that comes to us without obvious rational explanation; this probably involves the sensing capacity of the gut.

It is also the main emotional factory of the body. Most emotions at the chemical level consist of peptides, particular forms of proteins. Most peptides are produced in the gut lining. It is no coincidence that when I ask people to indicate where they experience feelings in their bodies, the majority point to their chests or their gut areas. (There are a few that point to their heads, but that is another story!) We can also relate to the correlation between emotional stress and illnesses of the digestive system and gut.

The heart and gut and the corresponding limbic and basal systems of the brain have been working together for a long,

long time. They operate at the subconscious level, and although we may previously have guessed or philosophised about their roles, it is only very recently that science is starting to tease out some of the secrets of this underworld. But many secrets remain untouched.

Cellular intelligence and the somatic network

Body intelligence does not stop with the triumvirate of head, heart, and gut. There are other smaller neural hubs across the body, and, ultimately, every cell is part of the total network. Each cell depends on its sensing capability to survive. Red corpuscles in the blood system need to be able to sense either nutrients or toxins in the bloodstream. Receptors in the cell membrane constantly vibrate as their mechanism for sensing their environment. When the cell senses a nutrient, it will resonate with it ('singing in harmony'), forming a connection which will cause the cell to open and allow the nutrient to gain entry, thereby nourishing the cell. If it senses a toxin, it will close up and attach itself to other cells in order to reduce the cell's surface area accessible to the toxin.

The body has a network of somatic cell markers which function together to sense their environment and to send information to the brain. This is both a neural network and an energetic system which enables the body to sense turbulence in its surrounding environment. For instance, an atmosphere of either tension or excitement will cause energetic ripples to start arising in our own personal energetic network. If further activated, these ripples will become waves, and, if unabated, the waves will become part of an inner storm. The way that this part of our intelligence system works is that the body does the sensing first. When the ripples initially appear, the somatic network will send signals to the brain as an early alarm system. The brain will then check it out in order to make sense of the experience. This making sense process will happen initially intuitively at the subconscious level, and then at the conscious level where we can use language internally to analyse and potentially explain the situation to ourselves.

We can see throughout these examples that the brain and body are integral components of an incredible intelligence system that operates both physically and energetically. They are partners in our amazing evolutionary journey.

In the next chapter, we will integrate both quantum and neuro insights to look at the phenomenon of human experience.

CHAPTER 4. THE HUMAN EXPERIENCE

While the human sciences have made huge progress in understanding the physical machinery of our brains and bodies, they still struggle to represent and examine the subjective human experience. While we can, for instance, examine the impact of emotional images on the brain, this still does not give us a reliable explanation of what it means on a personal level and what it feels like. What is the stuff of our own experiences? How do they register within us and sometimes rise to consciousness? What is the relationship between the mind and the brain?

These are big questions with significant implications for understanding the human experience. We have no concrete answers yet, but we are making progress. In this chapter, I will turn to both quantum theory and neuroscience to throw some light on this intriguing issue.

Perception

We should not underestimate the role of our brains in creating our experience. It goes far deeper than we typically think and is crucial to our hopes of living the quantum way. Let's start with visual processing.

Human sight is not simply a matter of transmitting external images through the eye which are then recognised in the brain. Quantum physics has already shown us that all matter we perceive is to some degree a collective illusion. As discussed earlier, we are surrounded by a universe of infinite energetic systems. Each cell in our body is a system, so is each organ and the total organism itself. Every material object is a system, including the book you are reading now. Items like a table appear still but they are not, at least not at the quantum level where there is perpetual motion within the atom as electrons orbit around the proton.

I realise that this is an incredibly difficult concept to accept. The key here is understanding that our brains have learned to represent these as objects in our minds in such a way that we recognise them as a tangible part of our everyday experience. And as all our brains are more than 99% the same, we will all agree that a table is a table. The problem is that we have always thought of these objects as existing independently of ourselves. But they do not. They materialise through interaction with human consciousness.

When we are born, we immediately start the process of building our own memories of the visual representations we have created in our own minds. These are known as internal models. They act as our constant visual reference base as we build our understanding of the world in which we live. If we do not have a direct or associated image of something, initially we are unable to see it.

There is a well-known documented history of how Spanish invaders of the Central Americas were able to get very close to the shores before the indigenous people raised the alarm. This was because the local people had no comparable mental images of big ships. They did not spend time pondering the horizon as they considered it to be the edge of the world. By the time they recognised the images, critical time had been lost.

The point is that even the act of recognition through seeing has to be learned. We don't always see exactly the same things. We build our visual memories by learning experientially and subconsciously where to put our attention. Our brains cannot possibly process every bit of information every second of the day as though it were new data. Subconscious memories are influenced by where our energy goes. Our attention and energy go hand in hand, allowing us to focus and remember the information that felt most important at the time. Images have meaning because we remember them as being important. If they have no meaning, they are likely to be bypassed.

In this way, the brain shows another insight into its sophistication. Every action in the brain uses energy and our supply of human energy, as we know from experience, is finite. So, this incredible machine in our heads is self-designed to use our energy to best effect, which means using the images we have processed before! When we look at each other we use more stored data from our memory than we do to process images as though they were new. In other words, when I look at you, I typically see 60% of you through previous images I have stored of you and only 40% of the you I see in front of me now!

Think predictive texting. Instead of requiring you to retype every word, the phone technology has built a library of many of the words you have used before. When it looks like you are about to process the same word again, this memory offers you a match which you can choose to reduce effort. The same principle applies to visual processing.

Our visual processing system is both complex and amazing, having built more and more layers of sophistication throughout our evolution. While there are many other species with specific visual capacities beyond our own, the richness of our visual imaging cannot be matched for its breadth of recognition, meaning and capacity to see the whole picture. We have 36 visual fields that we assemble into one meaningful whole. Our visual and emotional pathways are interlinked so that we can feel the power of evocative images. Our visual and language pathways are closely connected so that we can evoke imagery through the careful choice of the chosen phrase. When we put vison, emotion, and language together as evocative bedfellows, we unleash a cocktail of experience that is essentially and exclusively human.

We have focused so far on visual processing, but the same is true of our other senses. Sound does not exist outside of our heads, yet we live our lives as though it does. Sound is just the movement of external pressure waves which causes the bones in our inner ear to vibrate, sending signals into the brain which are converted into something we experience as sound. We never actually touch anything. What we experience as touch is the point where our own energetic system interacts with the energetic system of the person or object we are dealing with, where we detect a resistant force. Sensors in our skin relay this back to our brains where we conjure up the feeling of touch. Taste and smell (olfaction) are associated with the detection of chemical stimuli in our external environment that we convert into internal sensory experiences.

All these senses play a major role in our own personal blend of life, but they rarely operate in isolation; it is the combinations that matter most, the cocktails of sight and sound, taste and smell, touch and connection. Consider how the brain assembles sensory information in a synchronous way to create our meaningful experience, a process known as sensory binding. When a pianist presses the keys of his piano a whole range of sensory stimuli are triggered, including seeing the keyboard, the touch of each key and the sound generated. All these external stimuli travel at different speeds to the brain; so, if we just converted each as it arrived, the pianist would see first and then touch and sound would arrive at different later stages. This would bring a very different experience than that which the brain has perfected for us. The brain's capacity for integration and timing of processing signals means that it will first recognise the association between the signals and the need for them to be treated as an integrated package. It will then time the processing of these signals to give us one holistic experience where the pianist is able to create the impact her talent offers.

Ironically, while we have this wondrous human technology for seeing the physical world – with one important exception – we cannot see energy. The exception is light energy which we see every day. Yet, we can see only a very tiny portion of the energy spectrum, estimated at one trillionth! We cannot see radio waves, microwaves, ultraviolet, X-rays, and gamma waves, but they are bouncing around us every second. Think what a different universe we would be experiencing if we could see all of this.

Equally important is our experience of internal images. These are just as powerful as external images and our mind has its own internal energetic cinema screen, known as the retinoid stage, where it can project and experience them. Again, we totally underestimate how important our internal images are despite constantly referring to them. When we answer the phone, we immediately construct a mental image of the person we expect to speak to. Even if we don't know who is calling, we will conjure up a loose image which we associate with the voice, such as a man in a suit or a cold caller from a contact centre. These images set up a relational dynamic in the brain which helps us to communicate; we cannot engage with a vacuum, so a loose picture-fit will have to do.

The more we think about this concept, the more aware we can become of the internal images that drive our responses, a gift that can be very valuable when we try to understand our subconscious reactions in the moment. Such images are crucial in setting the tone for our behavioural responses and the decisions we make, and especially important in areas like personal confidence and performance.

There is an important link here to the way we construct our beliefs. Seeing is our primary sense, the one we trust more than any of the others. We use up to 60% of the cortex to process sight. Seeing and believing are inextricably linked. If we can see something our first reaction is to believe it. It has substance in our minds so we can easily relate to it. If we cannot see it, the opposite is true: we don't believe in it. The only thing that the brain cannot process is nothing: we cannot address a vacuum as there is no information to make sense of. So, we fill the gap with our own associated memories. But if we have no reference point to make sense of the phenomenon of invisible energy, we dismiss it. It becomes an inconvenience and waste of internal energy, so we behave as though it doesn't exist. This sits at the heart of why the energetic world has become so difficult for us to deal with.

Think of a rainbow. Many of us have seen a rainbow, something we may think just magically appears in the sky and then disappears again. Because we have seen it, we readily accept that it exists, and we do not struggle with its subsequent disappearance. This is an example of how all existing phenomena materialise. When the conditions are right, in this case where sunlight intersects with moisture in the atmosphere and we are viewing from the right angle, an energetic phenomenon becomes perceptible as something real. The same applies to the wider quantum spectrum where energetic phenomena become 'real' through human perception. Of course, while the rainbow disappears when the sunlight fades, our powers of visual perception stay constantly with us and reinforce our understanding of a tangible world.

Why does this stuff matter? Does it really affect our daily lives? After all, we have got this far without understanding it. Fair point, but here we need to make a fundamental choice. Large parts of our lives are built on assumptions about what is real and what is not. Science is now showing us that the boundaries that separate this polarised view of the world are invalid. So, we can choose to ignore the emerging evidence and stick within the confines of life as we have traditionally understood it; or we can recognise that, although initially unfamiliar, there is potentially a whole new dimension of understanding which can open us to lives we may not yet be able to conceive of. The former choice feels to me like being stuck in a cul-de-sac: it is familiar and comfortable, but it is taking us nowhere. The latter choice is not a quick fix, but maybe here we can take a lesson directly from the universe itself; to be exploratory, emergent, patient, and, above all, to keep moving forward.

Learning, language, and performance

Learning

The human brain never stops learning. When we are awake, most of our learning is going on at a subconscious level. At night when we are sleeping, we are integrating the memories of the day into our longer-term memory banks. A key biological mechanism for learning is brain plasticity, which is how we regenerate brain cells and create new connections and pathways. We process around 400 billion bits per second subconsciously and only 2,000 bits consciously; so, by far, most of our learning is done experientially rather than by choice.

Of course, we also need to be able to undertake conscious learning, whether it is part of our structured education or simply wishing to retain new knowledge or acquire new skills. This process is tougher and uses more energy. It requires us to use attentional focus. Remember that drained feeling when revising for exams?

Once the 'rules' or 'theory' of a new subject are established in the rational brain, to become skilled, we need to play with it. This is where the emotional brain also gets involved. This is not about getting emotional but rather about getting a feel for the skill we are attempting to learn. At this point, the brain operates like a guided missile system locking on to the target to learn by experiential adjustment. In this way, the cortex and limbic system collaborate to fine-tune the neural pathways which will master the skill; the cortex refines the rules of execution and the limbic system provides constant embodied feedback to perfect this. As these connections are tested and refined through practice, they become stronger and faster and the newly acquired skill becomes instinctive.

Think about learning to drive: starting with the mental effort of learning the rules and theory, moving into the clunky practical learning phase where we have to think consciously of every aspect of the new technique and practise it until it actually works, to the time when the skill is mastered and we can't even remember precisely how we drove home!

At the centre of this sophisticated learning is a very subtle process of prediction and error correction. We may not typically be aware of it, but the brain is always thinking ahead of our actions using visual images to predict the outcome and to keep our execution activity on track. Driving is again a great example of this. Our mind is constantly producing a stream of subliminal predictive imagery which keeps us on the road. This predictive flow, relying on access to memory, is much more energy-efficient than analysing all new data afresh. Yet, consider what happens when something unexpected happens when we are driving; we immediately react. So, despite using very little expensive conscious processing energy, this predictive system is perfectly attuned to spotting anything unexpected and instantly reacts to the change, doing so by matching the external images of what is unfolding ahead of us with the associated images held in memory. Any major difference leads to an instantaneous response. Another example of the magic of our brains.

As we get older, the learning becomes harder. This is largely because we have less available capacity in our densely interconnected brains. A child will learn naturally and freely because they are writing on a blank sheet of paper, while our older pages are already full of notes. So, the additional learning challenge for the mature brain is having to unlearn old habits, clearing out the stuff that is no longer needed. This requires the energetic surge of motivation and a sense of purpose. Without such energy, our adult brains will simply revert to the rules we have already written, and we become entrapped once again in our learned habits. We will revisit this challenge in later chapters.

Language

As our thinking brains evolved, we slowly learned to shape the previously animalistic noises we made into something we came to develop and recognise as language. Language became a powerful human force, a cultural accelerator which enabled communication and cooperation at very advanced levels. To truly appreciate this, it is useful to understand that this capacity for language developed over many thousands of years and came together through the integration of various parts across the human brain.

This is where the left and right sides of the brain come in. I previously highlighted the evolutionary architecture of the brain (basal, limbic, and cortex) with little reference to the left and right sides to offer clearer insights about the way each system works. When we look at the brain as a total system, the left and right side distinction is useful as it operates like a matrix with collaborative pathways running across systems and hemispheres. At the bottom of the two hemispheres is the corpus callosum, which is a neural bridge which handles communication between the two halves. Basically, the left side is focused more on rational processing and structuring information and is more attentive to the outside world, while the right side is more tuned into our inner world of emotion and intuition. So, think of it like this:

Left = rational

Right = emotional/intuitive

The left hemisphere looks outwardly and processes facts and evidence. It brings structure to our understanding of the world. Without this we would be swamped by data and emotion. The right hemisphere is more closely aligned with our limbic and basal systems, which enables it to be more intuitive and creative. Without this capability we would be robotic. The left is the classic domain of the scientist and the right belongs to the artist. We have the capability to be both.

There are two language centres in the brain forming the central pillars of the language system. Broca's area sits in the left hemisphere and is concerned with language structure and vocabulary, and Wernicke's area sits in the right hemisphere and focuses on meaning. Beyond that, the language pathways of the brain are intricately interconnected with both our visual processing and emotional systems. In this way, we have evolved this unique and powerful double-edged capacity: we can articulate our innermost feeling in a manner that can be shared with others through language, and language itself can be used to trigger the most evocative imagery and emotional responses in our inner world.

Furthermore, language is not limited to sharing with others; it is also used internally to shape our thoughts. Let's stop for a moment and try to reflect on our thoughts, more specifically the way we think. Firstly, to think about thoughts, we have to have a meta structure in our brain to think at different levels. Not all thoughts are conscious. As we discussed in the driving example above, we can process thoughts subconsciously when we are executing tasks, where we are simply exercising the existing neural pathways in the brain. Reflecting on our thoughts is another layer. Thoughts are registered in the cortex at the subconscious level in about 250 milliseconds, whereas it takes 420 milliseconds to register thoughts consciously in the prefrontal cortex, which is where we do our reflecting.

Secondly, can we think consciously without using language? Try it. It is very difficult. Language has become so integral to our conscious thinking that we cannot easily imagine what it would be like to think without using its structure. Yet, when the brain processes reflective thoughts, the language centres of the brain are at the end of the reaction chain and are inevitably 'the last to know'. So, what is happening before the language centres get involved? The answer is that we are primarily processing sensory impulses and internal images. We are far more reliant on our senses and visual memories than we realise. Language is involved only at the last stage when we want to raise the thought to consciousness, when its structure and rules give us something the thinking brain can relate to.

I find this fascinating. The implication is that we have already 'made sense' of the stimulus before we become aware of it. Language is used internally to represent meaning, not to form it. This has ramifications for our understanding of decision-making, which we shall examine shortly. What is important here is acknowledgement that the challenge of open-mindedness runs deeper than conscious reflection as this can only give is insight into the conclusions we have already reached. This undoubtedly has value and is to be absolutely encouraged to further our own growth and self-fulfilment, but the real gems sit at the subconscious level where the essence of our responses are formed. We will return to this in the next chapter.

Performance

The word 'performance' can have very wide connotations, so

I need to clarify what I wish to cover here. I am referring to our ability to be at our best in some experiential and typically observable form. This covers obvious performance areas like sport, but also art and expression. It is concerned with the ability to excel.

Adopted from the sporting world, being in the 'zone' refers to a mental place where we are totally focused on what we are trying to achieve and seemingly able to bring our best talent to the execution of the task. The phrase is used to denote someone being at their best performance level in the moment. Athletes and performers describe their experiences of the zone as stepping up to another level, a state where senses are heightened, and they lose any self-consciousness because they are totally immersed in the activity they have committed to. This is an observable phenomenon and spectators may comment how easy star athletes or performers make it look. This is because in physiological terms something very special is happening between body and mind.

In Chapter 3, I provided a broad outline of the wider human intelligence system: the 'zone' is a state where this total system is synchronised behind one goal. Particularly crucial here is the role of the heart as the source of our confidence. The heart (not the brain) transmits an electromagnetic wave across the brain and body which entrains the total intelligence system and causes it to act with one accord. This is a state of synchronisation. Such synchronisation is a state of inner energetic resonance where mind and body come together in harmony. Such inner harmony is not a normal state. Our bodies normally are in operating mode and all parts of the system will be undertaking their own tasks. So, what happens to bring us into the zone?

The trigger to confidence and to moving into the zone is a combination of familiarity and motivation. If the challenge we are confronted with is familiar, we will know what we are dealing with, and if our memory tells us that we have succeeded at this task before, we will expect to succeed again. This is why practice is so important. It not only helps us fine-tune the techniques of success, but it also reminds us that we can do the job. Constant repetition builds belief through hard-wiring the brain pathways needed to affect the outcome. When it comes to execution in the moment and in the glare of professional events, our emotions and tendency to overthink need to be kept to a minimum. We need to let our instincts take over. The execution machinery that resides in our basal subsystem is far more sophisticated than anything we can conjure up at a conscious level. Practice is about building the internal machinery and learning to trust ourselves in a given task.

This alone isn't enough; to stretch us, the challenge needs to be exciting and relate to our sense of purpose. So, if our purpose as an athlete is to excel in our particular event, the challenge will energise us because it is important to us and we will believe we can succeed because we have done so before. Of course, it is not that simple: our motivation can vary, and our focus can be undermined by fear and memories of when we did not succeed. I have worked with athletes who turn up to every game with the same talent as the last game, but they all know that their performance will also be reflected by the 'form' they feel in, which is essentially a temporary emotional state. In the zone, we channel our emotional energy as a positive force. When we are 'out of form' we instinctively and subconsciously call up anxious memories that undermine our ability to focus on the task at hand.

When we are confronted by a meaningful task that challenges us to excel, we are motivated and focused and in that moment nothing else matters. Physiologically the results are amazing and can be clearly tracked in laboratory conditions. HeartMath, an organisation based in the US, has done a lot of fascinating work on the heart-brain connection. When our intelligence system is entrained, we achieve energetic resonance at a frequency of around 0.1 herz. At this frequency, our power is amplified to a level which is potentially five or six times more powerful than our normal operating mode. This is not just about strength; it is about the energy we can generate for the same effort. Because the goal is clear and there is no emotional distraction, we can centre our resources on the challenge. This means that we can process information at much faster speeds than normal. We surrender to our instinctive powers. We are at one with our environment. This is a state of flow, one of the critical quantum principles described earlier.

So, does the need for familiarity mean that we can never be confident about a new challenge? Fortunately, this isn't the case. Here is where our associative memory kicks in. As long as we can sense a similar association in memory that matches the challenge, we will already feel we are in with a chance. For instance, a footballer may instinctively feel that they could be good at hockey. They will know that they are not the same thing and they are not initially likely to succeed to the same level, but they have no fear of the challenge because they know they have the motor skills, balance, and broad tactical understanding of the challenge to do well.

This translates well beyond sport. Artistic expression is a state of flow; the artists immerse themselves in the expression of the inner image and sensations that have triggered the inspiration. The 'zone' for the artist is inner access to a richer and deeper tapestry of sensory information which heightens their sensibilities. Someone who is an experienced and successful presenter will be able to take on a wide variety of topics because they know how to engage an audience and get them on their side. Of course, there is a limit, but those limits are much wider than for someone who only feels comfortable in presenting specialist material. There is an important distinction here between trusting your material and trusting yourself.

However, the zone is not a sustainable state. We can access it on a temporary basis, but the intelligence centres of the brain and body have their day jobs to return to, our default operating state. In fact, research shows that sustaining high levels of performance depends on the ability to effectively step in and out of the zone. This is important. None of us is constantly 'in play'. We need to use our zone time wisely and sparingly. Equally, we need to use our downtime sensibly, allowing enough self-care and energy regeneration time to ensure the fuel tank is fully topped up and ready to go. We may be able to drive a Ferrari in the zone but even the fastest cars run out of fuel.

Think this through carefully. When are you really 'in play'? These are the crucial times that you need to be at your best, which is not the time for practising or experimentation. Yet, if we do not give ourselves exploratory time, we cannot develop our performance horizons further; we cannot grow. So, what is your personal strategy for getting the most out of your talent and passion? You need to be aware when you are on the training ground, when you are restoring your energy, and when you have to be at your best. If you think you can constantly operate at the top of your game, you are fooling yourself.

This takes us into the world of resilience and belief. We will revisit this subject again later, but for now we can confidently declare belief to be a prerequisite for resilience. There is no resilience without belief. It is belief in a purpose that inspires us to get up again when we are knocked down, that allows us to see beyond the immediate setback and onto the horizon that awaits. In the same way, our ability to sustain effort without a relatable belief is extremely limited.

The human chemical cocktails which underpin performance offer us a flavour of what we all need to succeed. Hormones are the chemical messengers of the body. They are generated and secreted by the endocrine system, that is by our glands which are located across the body and within the brain (brain hormones are called neurotransmitters). As explained earlier, neurotransmitters act as a connection junction enabling cells to activate specific pathways, triggering a particular response.

The chemical prerequisite for any form of peak performance is adrenaline, the essential priming agent and energy fuel of the body. Adrenaline is triggered when we instinctively connect with a purpose or challenge. This means that we are energised and excited. But excitement alone is not a basis for higher performance; in fact, if untethered it becomes a distraction. We need to focus and channel this energy. The key hormone to do this is acetylcholine, which balances our adrenaline and helps us to stay calm. Calmness is vital to enable us to have instantaneous access to all our senses so that they can be recruited in the cause of our performance. The third hormone worthy of special mention here is dopamine. Dopamine is typically known as the reward hormone. It is the chemical leader in keeping us on track towards achieving a goal. While we all need the inspiration of a higher purpose to give our lives direction, we also need short-term dopamine hits to give us the everyday belief that our effort is worthwhile. Dopamine underpins our sense of achievement when we knock out those necessary and immediate tasks that so often clutter our days.

Any top performer knows that having the right state of mind at the right time is a prerequisite to success. Equally it is our state of mind that can block us, destroy us, and condemn us to failure. It is a potent mix of opportunity and threat. Often, we try to address our state of mind challenges through another mind-based process, such as by trying to think ourselves into a better state. It will not work. We cannot balance mind with mind. We can only balance mind with body, which means tapping into the subconscious sensory data and mechanisms that sit across our bodies. These serve to ground us in the reality of our experience in the moment, rather than some mental construct we are trying to impose from a rational space, and which denies the honesty of our emotions. We cannot think ourselves into confidence; we have to feel it. Thoughts can lie but emotions cannot.

Meditation is an excellent example of a technique for grounding us in the reality of the moment. When we surrender to a meditative practice, usually by focusing on our breathing, we calm the overthinking of the rational brain and instead allow ourselves to access our deeper intuitive intelligence. This helps us to anchor ourselves in a place we know, a place of belief, protected from the incessant enquiries and anxieties of the racing mind. This facilitates focus, a key component of performance execution.

However, I am aware that many people struggle with meditation as a structured process. In this case, there may be many other natural practices that will support you to calm the mind, including running, walking, swimming – basically, any rhythmic activity that does not require analytical thought. The body knows how to calm itself; we just need to get our tendency to overanalyse out of the way, clearing the ground for the soothing heart to take control. Visualisation techniques can also be very effective in this way.

CHAPTER 5. SELF-DISCOVERY

The self

Now we come to the self. Understanding who and what we are is surely essential to our ongoing wellbeing and our opportunity to pursue happiness. If we misunderstand our essence, how can we navigate our way to a fulfilling life. If we cannot understand ourselves, how can we understand others? If we are ignorant of our core inner dynamics, every personal decision we make runs the risk of focusing on a false premise. And if we collectively fail to nurture our deeper needs and aspirations, any society or organisation we build will fall short of its true potential.

The self is a notion that we have been educated in the West to value highly, arguably more than anything else. It is usually represented as a unique personal mind state that is our essence. We look out onto the world from the self and navigate our way through life. It is the only constant in our lives. Such a notion has been a cornerstone of psychology throughout the 20th century. This thinking is so embedded in us that we take it for granted. But now science is throwing up some very important challenges to this assumption.

Firstly, there is no evidence of a constant self in the brain. Although we recognise the prefrontal cortex as the region which plays a significant role in self-awareness, it cannot be pinned down to one specific neural centre. So, is the mind the self – are they one and the same thing? A good question. I suspect a lot of people would answer affirmatively. But then there is the complication of deciding if we mean mind as a whole or just the conscious mind. My own interpretation is that most people associate the self with conscious thoughts. But conscious processing is a tiny proportion of what our brain handles. We process 40 million impulses per second subconsciously and only 40 consciously. Ninety-five percent of our behaviour is triggered by subconscious programming and only 5% consciously.

So, how can the conscious mind be self-aware if it has so little access to what is going on? And if we then argue that the self must be the whole mind, by definition we are saying that is 95% + unaware. This does not stack up with any meaningful idea of reliable self-awareness.

I don't want to get too heavy here, but if we take ourselves back to the quantum principles discussed earlier, we will see the brain and mind as interdependent. One cannot exist without the other. The source substance of the universe is energy, and our minds are energetic creations anchored in the physical world by our brains. Energy creates the brain and the brain in turn creates the mind, and they shape each other throughout our lives – a perfect illustration of the entanglement of the energetic and the physical.

Neuroscientists are more likely to see the self as a stream of consciousness, which changes every second of the day as we process more experiential information. It is akin to a central energetic force that holds together our memory and outlook. Similarly, the quantum physicist could refer to the self as a field of meaning, a state of ongoing resonance that binds together its own energies.

In parallel, our understanding of the quantum realm shows us that nothing exists in isolation. The implication is that our selves do not exist independently of the world they live in. There is no negotiation between the independent entities of self and environment, rather an ongoing process of mutual shaping. The self is then a product of this process, not something that sits in any way apart from it. This means the self was not born as a new identity; rather it was born as potential and is shaped from the very beginning by its environment and life experience.

This is a very unfamiliar perspective on the self and one that may intuitively cause some discomfort as it is challenging us to give up our own deeply held sense of who we are. Yet, I want to offer you some reassurance. My own experience is that potentially the reverse is true: freeing ourselves of some of the old concepts of self can evoke a renewed sense of freedom and choice. We do not need to hang on to many of the beliefs and perspectives that stop us from moving forward.

So, let's explore our sense of self in more depth. Why do we have such a need and how do we sustain it? Some of the beliefs we explore will stand up to scrutiny, but others may be less reliable. In turn, we will explore our sense of feeling unique: our own story, our sense of embodiment, our need for social acknowledgement, and the hot topic of free will.

Feeling unique

Let's start with our sense of being unique. Despite challenging the idea of an independent and constant self, I have no doubt that each and every one of us is unique. While our brains are 99% the same in operating design, the connectivity of each brain develops according to its genetic template and life experience. It is this 1% that makes each one of us uniquely who we are. Not only is our appearance unique, but so is our world of inner experience. In understanding the infinite potential of our brains and our energy, I am readily assured that there is room in the universe for every one of us. So far, so good.

Our own story

Secondly, we have our own story, a sense of continuity which evolves through life. It is self-evident that memory plays a huge role in this. We refer to memory to make sense of our everyday lives and the lessons we have previously learned about how to think, feel, and act. We have built our identity through memory and actively integrate new experiences every time we sleep. In fact, it could be argued that we are our memories projected into the future. Our memory is our information base from which we interpret everything that is going on around us, both consciously and subconsciously. We don't make it up as we go along; we constantly refer to memory to interpret stimuli from the outside world and then we respond accordingly. No memory means no response. The bottom line is that memory is not just something we refer to when we are trying to recall something consciously; we are using it to guide us every second of the day.

So, if this is how it works, what is the problem? The problem is the story we tell ourselves. We have a deep psychological need to have our own story make sense and support our current self-identity, both for how we express this identity to others as well as ourselves. We cannot possibly present an objective and comprehensive view of everything we have experienced and learned, so we select the memories that help us feel congruent in who we are now. It is like looking back up the river we have travelled and making sense of how we got here, but 'here' is simply another experience along the way. When travelling down the river we are going with the natural flow, and our journey unfolds largely outside of our control. But we have been told about the importance of the individual and understanding the self – the mind-centric phenomenon from which we view our lives. We need our story to make sense of our journey and to give us a position from which to anticipate the future.

This process of memory selection is primarily energetic. Memory works essentially through resonance, and the memories we most easily recall are those with the highest 'energy tag': the stronger the original experience, the greater the 'tag'. This enables that memory to stand out among the noise and get noticed. We then select these energetic peaks from our current vantage point and relate them associatively into a story, the 'story of our lives'. This is an emotional need and happens both consciously and subconsciously. We typically strive to create the story that best supports us now, but becoming overly attached to it will cause us difficulties.

When our stories become unduly scripted, they will serve as a constraint to our freedom to flow. If we have anchored a habit to tell ourselves who we are, any challenge to the authenticity of that story can create fear, and we may respond by digging ourselves even deeper into the trench we have already created. We will fight for this identity and potentially shut ourselves away from new information and experiences that do not support our story. This is less about a story for telling others but, more importantly, one for ourselves. It can therefore be very disturbing to have this sense of continuity threatened. But is there really anything to fear? By holding on to our story we are hooking ourselves to the past. We cannot be now what we were in the past, even one second ago. We constantly evolve through the dynamics of our brain and our stream of consciousness; herein lies the amazing freedom of letting the self go.

The challenge of self-discovery runs through all quantum learning and calls us to rethink who we are at the deepest level. By doing this, we open ourselves up to a universal perspective that allows us to accelerate the next stage of our intellectual and experiential evolution. I invite you to ask yourself this question: what am I telling myself? We will look at this further in Chapter 7 when we look more closely at beliefs.

We must also recognise that we do not have control of our memories. We all carry the experiences of relative trauma and disappointments that have been a core constituent of our lives. We would not necessarily choose to carry this with us and the associated pain and fear that we can still feel. Yet these memories stay with us to protect us from similar situations that may arise in the future. The problem is that in the understandable struggle to avoid revisiting such pain, the story we tell ourselves becomes false. We construct a self-identity built partly on denial and imaginary escape from our past. But it will not work. We are very clear now that treatment of all trauma is best served by revisiting the source of the fear, albeit with skilled and sensitive support.

Turning your back on danger may feel like the only escape route in the moment, but when the danger follows you it will only make you even more vulnerable. Consequently, many people experience a great deal of stress in their ongoing life battle to get away from their past. The same can be said of hanging on to a rigid view of ourselves.

Sense of embodiment

Every one of us has a sense of being in our own body. This is so obvious that we rarely think about it. Yet, this involves a sophisticated system of inner and outer intelligence which can differentiate between what is going on inside and what is happening outside our bodies. Because of our fixation with the physical world, we assume that our own boundary to the external world is our skin. In physical terms this may be right, but it is not true in holistic or quantum terms. Each of us transmits an electromagnetic field which extends typically around five feet outside of our body. This energetic field is transmitted primarily by the heart and is just as much part of us as our physical properties. It is an integral part of our ability to sense and interact with our environment.

Our sense of body comes from a series of intelligence systems that allows us to map our bodies in our brains and to sense our location. If we were able to see the appropriate parts highlighted at the top of the cortex, we would see a representation of our body.

We would recognise our limbs, face, shoulders, and so on, although the picture would look more like abstract art than traditional portraiture. This is because the area of representation will reflect the nerve density of the body part. In this way, we see large hands and feet, tongues, and sexual zones. Additionally, we have parts of the brain that enable us to sense our location: this includes the hippocampus, which has a primary role in memory and also doubles up as our own GPS, locating us in our environment.

Mirror neurons are located throughout our brain and enable us to feel what others feel. As the name suggests, they mirror what is going on in others so that we can recreate that experience within ourselves. This again relies on energetic sensing which then triggers a biochemical reaction within us. This is the basis of empathy and emotional contagion. Interestingly, the same neurons 'also turn inwards' to give us our own sense of selves as part of the process of bringing emotional experiences into our awareness. What I find especially fascinating is the way we then deal with differentiating between our own and others' emotions. As our mirror neurons are activated by the feelings of others, sensors in our skin send parallel signals to the brain indicating that 'this isn't me'. These 'null signals' nullify the intensity of the emotion. In this way, we can interact very effectively with the outside energies without getting entirely swamped by them.

Social acknowledgement

The emotional system of the brain (the limbic system) is designed for connection with others. Despite what we sometimes like to think, it does matter to us what others think and feel about us. In fact, other people are the biggest shapers in the development of our brains. The emotional template of the brain is laid down in the first three years of our lives, and our primary caregivers (usually our parents) are the ones who will be informing us both consciously and subconsciously what we should be looking out for and how we respond to it.

While we talk quite generally about feelings in our daily lives, the neuroscience community typically recognises certain core emotions which can be recognised as the primary foundations, from which all other feelings are blended. These core emotions are often recognised as fear, anger, sadness, disgust, shame, happiness, and love. Of course, it is difficult to agree on such a list in absolute terms since the language used will have different associations for all of us. Nevertheless, we can readily recognise the positive connective power of love and happiness. But nature has us both ways, presenting us with both stick and carrot. Shame is nature's way of ensuring that there is a further deep connection to the family or community of which we are a part. It is the stick rather than the carrot. Like all emotions, shame is purposeful. If our attitude to group loyalty was flaky, we would not have survived as mammals. Primates who are cast out from their groups often die of shame, and in some human cultures we still see the honour of the family as being sacred, transgressors sometimes paying with their lives.

Likewise, we constantly seek validation from others. We may be good at hiding it, even from ourselves, but we have a deeply rooted need to belong, to follow, and sometimes to lead. Therefore, just like our motor car driving example discussed earlier, our emotional system will operate like a missile guidance system constantly locking onto our emotional targets and subconsciously adjusting to stay on track with who we think we want to be. This ongoing search for validation can be a great source of assurance to us, but it can also get us tangled up in an emotional web of disappointment and deceit if we are chasing the wrong source of validation, one which takes us away from who we really are.

This strikes at the heart of this book. What emotional assurance are we chasing and why? Surely, with a little reflection, we can all identify many instances where we are emotionally tied to something or someone that really hurts us, yet we continue to feed the addiction. Furthermore, it may also be true that we hold on to a view of ourselves that can no longer serve us. Subconsciously we have become a willing partner in our own demise. The choices of emotional alignment and acknowledgement are both subtle and powerful, and hugely influenced by our life experience captured in memory. Any attempt to free ourselves from this grip will take courage and inner wisdom.

Free will: a sense of agency

Lastly, we need to examine our sense of free will. This highly valued aspect of our lives gives us a sense of personal conscious choice and ownership; the freedom to make our own decisions about where we use our energy and the outcomes we choose to pursue. Indeed, without this who are we? Clones? Robots? This can be a highly sensitive subject area, and the good news is I have no basis for challenging our individual uniqueness. Like anyone reading this book, I have a sense of making my own choices about my life, ultimately unconstrained by others who may influence me, but the buck stops with me. I can choose to heed others or ignore them, a clear illustration of free will. Nevertheless, there is a level at which this can be challenged.

Let us look at decision-making. We are encouraged to believe that, particularly in the business environment, we make fact-based conscious decisions based on a balanced view of the evidence available. Corporate businesses have made an entire subindustry of decision-making, gathering an entourage of fact gatherers, business planners, and accountants. Number crunching has become the necessary evil of business life, the path to progress defined by the tracks of rational thought, analysis, and projection. So, this is how we make good decisions, right? Wrong!

Neuroscience research shows us that 90% plus of our decisions are made at an unconscious level and this means that they are driven by our instincts and our emotions.

Surely boardroom decisions are evidence-driven? No, they are not; and here I draw not only on neuroscience but also on the many years I spent either submitting or evaluating business cases for board approval. Let's think about such a decision-making process on both a science and business level.

Our first response to any scenario we are presented with is instinctive, the fastest part of the brain that is intrinsically intertwined with memory. Typically, we match the presented idea or stimulus with an internal image that we recall from memory, and this image acts as an instantaneous catalyst that triggers our internal response. Virtually simultaneously, our instincts trigger our emotional response. This is where the amygdala attaches an emotional value to the idea, giving us a sense of how important it is or is not. Within seconds we will instinctively feel positive, negative, or neutral to the idea. If the response is positive or negative, this is not easily turned around later in the process, although it is possible. If it is neutral, the instinctive brain disengages.

Late to the party, the rational brain arrives, but the die is already cast. The influence of our instincts and emotions means that we will now look for the evidence to support the position we have already taken. All supposed facts are processed with an emotional value, and it is this system which decides whether or not they are acknowledged, valued, and influential.

If anyone is presenting a proposal or selling any idea to anyone, they have to win the 'hearts and minds' battle first. This means they have to create a positive emotional connection with their audience and then they have to present them a vision of success that appeals to them (the instinctive system is highly visual). The so-called facts cannot win the battle for them. Rational data is reassuring and provides clarity for purposes of execution planning, and to allow testing of the idea, but it always comes late in the process.

I recall in my corporate career having submitted a proposal to the group company board in the Netherlands. I had built the business case and my Netherlands-based boss was presenting the argument. To me it was a 'no-brainer', so I was devastated when I heard the proposal had not been accepted. My response was to jump straight on a plane and go to the same board to present the argument myself. This time they approved it: the facts and figures had not changed, but they felt my passion and met the guy who would be personally responsible for the success of the submitted plan. They bought me and my belief; the business case was simply an ongoing monitoring check. People buy people!

So, what has this got to do with free will? Everything. The same process applies to any decision we make about our lives. Remember, we process 40 million impulses per second subconsciously and only 40 consciously, so our subconscious mind is one million times faster. What chance does the conscious mind have of controlling such a torrent of information flow? And what drives our subconscious mind? The answer is memory. While we like to think we make conscious choices about our life, in most cases consciousness is simply the destination of the internal process, the point at which we become aware of what we have already decided. We convince ourselves that we have freedom to choose at this point, but it is largely an illusion created to give us a sense of agency and of being in control.

In this way, are we not pre-programmed? If our response to the situations we face is largely driven by memory, what is the point of the conscious mind? This is an amazing question! (I am glad I thought of it!) In all of my study to date, very little impresses me as much as the force and wisdom of evolution. As human beings we have evolved the prefrontal cortex as the seat of our self-awareness and our conscious mind. It is the fastest evolving part of our brains. Why? Why is self-awareness so important to us? While it may currently have a fledgling role, it nevertheless offers us the potential for greater authentic, conscious choice; a potential that can only be unlocked if we really understand the way we work now.

As humans we have learned that slowing down our responses gives us choice. In evolving from instincts and emotions to thought, we learned that more response time gave us more options and therein the capacity to choose. This truth is as valid now as it ever was. Unfolding before us is the opportunity to raise our consciousness beyond the level of physical perception and the prejudices of our own memories. By understanding our own inner dynamics, we can learn to nurture them towards a better future, however defined, harnessed through our higher consciousness. We may not have harnessed our subconscious world yet, but we are at least shining a light on its potential. And to the question, 'Do we really have free will?', I would offer the answer, 'Not yet!'

The mosaic

In the midst of this apparent attack on the sanctity of our

sense of self, I would like to offer a very simple, constructive, and liberating model. I believe we have generally become far too attached to our own egocentric views of ourselves and these views are too rigid. We are not our thoughts, nor our feelings, nor our instincts. We are not just our memories nor our projections into the future. None of these phenomena on their own can label us or accurately represent us in our entirety. Our essence is the whole, the energetic force that holds us together as one field of meaning. It is constantly moving, accumulating experience, and looking in on its own journey to be able to predict where it should go next.

The freedom comes from letting go of the self-protection and self-judgement that we carry so heavily on our shoulders. We have not chosen our subconscious responses, but rather we have learned them within a realm that we have hardly understood. Each of us is our own dynamic mosaic of parts which together comprise the story of our lives. We can educate ourselves to reflect on these parts and to understand why some parts are triggered into action in some situations and some are not. We can learn to be kind to ourselves in our search for understanding and to recognise that judgement is a barrier to insight. It is in our role of self-observer that we can avoid the pain of surrendering our essence to any one part of the whole. We are none of the parts, but we are all of them.

Surely our most fruitful task now is to continue the process of understanding the unknown aspects of ourselves, to shine a light into the shadows which have hidden us away for so long. In Chapter 8, we will look at our unfolding understanding of these unknowns and encounter the realms of belief and spirituality, including an exploration of what we might mean by the human 'soul'.

CHAPTER 6. THE QUANTUM ORGANISATION

In this next chapter, I will bring quantum insights back into a more tangible context, that of the organisation. The majority of people experience organisations as part of their everyday lives, either interacting with them socially, economically, or politically, belonging to them or being employed by them. They are a common feature of our existence. Equally, a vast majority of these organisations are no longer fit for purpose. The principles and thinking we used to design and build them are outdated and languish far behind the rapidly emerging understanding of human needs and what best enables us to perform at both an individual and collective level.

The concept of a quantum organisation is new. A number organisations have explored working with such principles, but they are still very few and the thinking is accelerating.

My own organisational insights stem from 30 years of senior management roles in corporate businesses plus a further 15 years coaching leaders and top teams. I have been fortunate to have roles which have allowed me to travel the globe and experience many cultures. In particular, the more recent years have exposed me to new emerging entrepreneurial organisations which are naturally and intuitively seeking better ways to lead and organise their people.

The days of the traditional corporate organisation are rapidly disappearing, a phenomenon recognised by many leaders facing the struggles of attracting, retaining, and sustaining the decreasing pool of talent who are prepared to make themselves available to the corporate employer. Emerging generations are looking for purpose and meaning in their lives and are not responding positively to the soulless corporate model, which continues to treat people like factory units. But corporate leaders are struggling to find the answers to this dilemma. Old thinking with minor adjustments will take them around the same loops back to the current crisis. Incremental change is no longer enough. Change has to be fundamental, radical, and come from the heart.

So, let's understand how we have got to this organisational place and how we get out of it. We can point to many commercially successful organisations, but my argument is that they have succeeded despite their organisational approach not because of them, and, as new models emerge, they will be left behind. Furthermore, the human and environmental price paid for this success has been too high and is no longer sustainable.

This analysis is not about judgement. I will not point the finger at current leadership regimes as their behaviour is simply a reflection of what they themselves have learned in the past. They grew up in traditional organisations; this was the environment their brains were exposed to, where their personal strategies to survive and thrive were honed. But now the demands of our organisational and natural environments have moved on at an accelerated rate. Even the most dogmatic traditionalist can see the strains in modern global society and the damage to our natural habitat. Organisations cannot operate in a vacuum, and the crises of environment and consumerism challenge us all to look again at the way we perceive the world. In this way, while my focus in this section of the book will be organisational, it is entirely consistent with an emergent world view where humanity itself has achieved a new level of self-understanding which allows it to build a significantly more informed, safe, sustainable, and fulfilling future.

Likewise, organisations of the future will need to have the courage and will to nurture and channel this quest for meaning and purpose. There can be no 'no-go' areas for discussion. Older generations were told to leave their emotions at home and bring only their professional selves to work, resulting in a complete polarity of our experience between home and work. This so-called 'work-life balance' issue is a complete mental fabrication, initially encouraged by those who simply wanted us to behave predictably and do as we were told. The reality is that we cannot just leave our emotions at home, and we cannot just leave our work experiences at work. Our emotions sit at the heart of who we are and cannot be ignored.

Overarching principles

There are three overarching principles which I believe we need to understand and embrace when taking on the challenge of creating a truly quantum organisation, which are summarised below.

Energy

First and foremost, a quantum organisation is an energetic entity, a collective state of being. Leaders of such organisations consciously manage the organisation as an energetic force, aware of the needs of its people and how to optimise this potential. The success of a quantum organisation is based on energetic flow. It does not aim to box people up into controllable units. Instead, it is designed to channel natural energy into the unfolding purpose of the total business.

A belief in people

A quantum organisation represents an absolute belief in people; a people-centric operation that aims to create an environment that will enable people to get the best out of themselves, both personally and collectively. The organisation is built to support the needs of the people, not the other way around, as is currently the case with many traditional corporate businesses.

Connection

Alignment of activity comes through energetic connection – that is commitment to a shared purpose which will create the opportunity for both individual and organisational growth and sustainability. Ongoing connection to this purpose is vital and signifies a critical role for effective and meaningful communication across the business.

We will deal with these features respectively in terms of culture, structure, and alignment. We will then take a look at what this means for quantum leadership. All of these features need to be understood as pieces of an interlocking jigsaw, and what matters most is the total picture (see Figure 2).

THE Q9
FEATURES OF A QUANTUM ORGANISATION

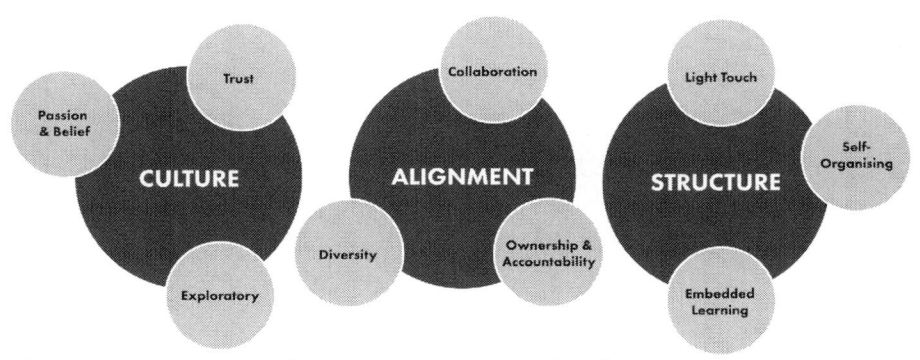

Figure 2: Features of a Quantum Organisation

Cultural features

In this context we can think of culture as the energy of an organisation, the unseen realm of feelings, instincts, and thoughts which cannot be tied to an organisation chart. Do not underestimate its power. As Peter Drucker famously said, "Culture eats strategy for breakfast!" Strategy is important because it channels the energy of a business, but it is energy which gets things done. No energy equals no progress. Having an abundance of aligned energy is what makes great businesses and organisations.

We can also see this in an evolutionary sense. Human evolution benefited enormously from certain cultural accelerators which created a platform for a collective momentum that pulled us well away from other species. The classic example is language. Between us we created a medium for far more advanced levels of communication and shared understanding. Language became a natural collective force that spurred us forward at hitherto unseen speeds. The same principle applies in any culture. Can we learn to devise and nurture our own cultural accelerators to open the path to rapidly advanced performance?

We now need to break these down further into nine critical features to help to translate these principles into tangible strategies that can be used to move organisations in a quantum direction. I have settled on nine features. I am not wedded to these nine as the only elements that matter (nor even their names – you may well have a slightly different term for the same concept), but I think they are an excellent starting point. When I work with clients, I am very keen to ensure they understand the overarching principles, but I also want to ensure they take ownership of their own journey. You will see later in this chapter, where we look at the early experience of The Happiness Index in their quantum journey, that their 'Q9' features vary marginally from mine. This is entirely consistent with choosing the language which resonates most effectively for your own organisation.

Trust

Trust is an absolute prerequisite for top collective performance. In so many organisations we have become accustomed to undercurrents of distrust, either of bosses or colleagues who are largely in it for themselves. It is an implicit law of the jungle played out as office politics. Like so many bad practices, it has become normalised as though there is no other way. But there is.

As I explained in Chapter 3, the emotional system of the brain is designed to engage with others to create communities and social structures. Engagement is a natural state of flow for us as long as we feel physically and emotionally safe. The key hormone of engagement is oxytocin. It opens us up at all levels, from the cellular to the total organism. When oxytocin flows, we want to become a part of something bigger; we release our otherwise tight protective grip on ourselves, and we become more attuned to our environment. Our nervous system is balanced, our heart rhythm is coherent, our eyes sparkle, and our capacity to focus and to energetically sense others is enhanced. Our minds open up and we are ready for learning. This is an instinctive state where we are in thrive mode, and we are activated for growth.

Contrast this with the state of distrust where our survival instincts take over. If we sense danger, including emotional threat, we withdraw to a place of inner safety. We may play the game of compliance because we want to keep our jobs, but we will now only offer the bare minimum in performance and hold back all our hidden potential for a better cause. As explained earlier, when the brain is in survival mode, we lose perspective, objectivity, and creativity and conserve our resources instead for entirely self-protection purposes.

Organisational values sit within this gambit of trust. I have been consistently disappointed over the years by the 'tick-box' approach to values in most organisations. It hardly matters what values are articulated on the company notice board or in the latest PowerPoint presentation; they are largely ignored anyway. What matters is how people feel. Shared values should be a point of emotional connection between people.

The human brain builds a library of emotional values throughout life. When we explored visual processing in Chapter 4, we saw how internal models are held in memory and used to reduce the need for constantly reprocessing new images. The same principle applies here. It would be far too energetically expensive to process every emotional stimulus afresh. Instead, we create these building blocks in subconscious emotional memory and use them to enable us to have an instantaneous response to emotional triggers, to immediately recognise what is morally right from morally wrong, what is fair and what is unfair. Shared values are the emotional bond that create a community. When we are aligned through shared values we are in a state of energetic resonance, and we amplify the energy collectively available to us. They become the fuel of our intent.

People who don't share similar values will struggle to trust each other and will not create an effective community. Building trust in an organisation is therefore fundamental, not a 'nice-to-have'. Any organisation with quantum ambitions must invest in achieving high levels of trust among its people, both in the leadership and among themselves. It must carefully observe and monitor trust and be very clear what it is looking for and what interventions are appropriate. Our bodies have a highly sophisticated system of internal biofeedback, which allows the body to sense when we are in balance, our optimal state, or if we are out of balance and how this is occurring. Organisations need to follow this lead and use both technology and their own senses to continually feel the pulse of their business. In the quantum world, there is no fallback to hierarchical authority. It is about having developed the knowledge and skill to invite people into a world where they can thrive and offer their very best.

Exploratory
It is a natural human state for us to be curious. Evolution has given us the instinct to explore the next boundary. If we had simply settled for what we had, we would have stayed there. This started right down at the cellular level. Single-celled amoebas originally learned to move in order to chase nutrients in their environment. Its chances of survival were very limited if it waited for food to come to it. As humans evolved, their curiosity extended beyond their communities and immediate location, and eventually *Homo sapiens* moved beyond the continents of their birth. Roll forward to today, and we are still fascinated by what lies beyond, the open sky and the greater universe.

The human concept of exploration links directly to the emergent nature of the universe itself. The universe has unfolded over billions of years, establishing energetic systems along the way. There is no evidence of a grand plan where the shape of the universe was preordained. This translates to human nature itself. While we need a sense of direction, nothing is ever quite as planned. The emergent approach is then to stay connected with our sense of purpose and the direction it takes us; the practicalities of the journey will unfold along the way. It is about having sufficient clarity of vision to press on, but, at the same time, being open to the experiential learning that will come along the way. Plans do not have to be perfect. In practice this means being purposeful, flexible in approach, and open to everyday learning.

Just like evolution itself, a quantum culture needs to be exploratory in outlook, valuing the learning opportunities that go with experimentation and creativity. If we are denied these opportunities for human expression, we will stagnate. Evolution never accepts failure. If one road to progress is blocked, it will find another. We have evolved through creating new connections and developing new patterns to help us achieve our goals. We need a culture that supports us to feel brave enough to step into the unknown, recognising that mystery itself is just another opportunity to extend our knowledge. We need to be allowed to make mistakes without being harshly judged, providing we learn from the experience and bring that learning value back to the organisation.

This commitment to exploration also applies at the personal level. The quantum individual is someone who understands the need for self-discovery and openly welcomes the journey involved. They are someone who learns to sit comfortably within their own skin, recognising the importance of self-reflection and valuing their trusted colleagues as mutual learners and providers of interpersonal feedback. This can and should be a very rewarding ongoing experience.

Passion and belief

Passion is triggered by vision and purpose, when we can see where we are going, and we feel good about why we are going there. It is an energetic force that can lift us all to higher achievement. While trust is a calm energetic state, passion is highly energised and contagious. It stirs our energy and mobilises us to move forward. Passion and inspiration run hand in hand. When we are inspired, we can see an opportunity to thrive, usually one that is just out of grasp but is there for the taking. We feel the need to act and to seize the moment. To fulfil its potential, an organisation needs a vibrant culture, one which is energised by a passionate commitment to its purpose.

There is a rich hormonal cocktail underpinning passion, including adrenalin, oxytocin, and dopamine. Adrenaline fires us up for action, oxytocin engages us, and dopamine rewards us for our effort and keeps us on track. As human beings, we need the inspiration of the big picture vision to give us that feeling of being part of something greater: we also need the short-term rewards, such as the dopamine hits that tell us we are making progress.

Such high energy states are not permanent. We need to keep rebalancing our energy to refill the fuel tank. Such voluntary bursts of energy are invigorating, especially at critical times when extra effort is required. They not only inspire us in the moment, but they also keep us ticking over after the initial rush has subsided. When we have experienced how extraordinary these moments are, it gives us belief and we want more. We have felt our own hidden potential and we will take that feeling with us. It is a bit like playing golf. I am a poor golfer, but I love playing the occasional game with my son and with friends. Even though most of my shots are wayward, there are always just a few that are perfect (all golfers will recognise this), and at those moments I can see what I am capable of and I experience what is possible. It is inspiring, and it keeps me coming back for more.

Organisations need to take very seriously this need to inspire passion in their people. It has obvious implications for their approach to leadership and for the leaders they appoint. It encompasses both key events and everyday operations. Key events are the time for establishing direct shared connection with the cause people feel passionate about. Everyday activity needs to ensure that there is sufficient ongoing connection with that passion to keep it alive and breathing.

Structural features

The quantum way has radical implications for the way organisations are structured. This is the point where many existing CEOs will need to take a deep breath and push through. Internal struggle itself can be a very good thing in that the brain is making room for new learning, for creating new pathways. While I am sticking to quantum principles here, each of these principles is underpinned by my own experience as the CEO of a corporate business. Please don't evaluate any one of the features in isolation from the others. The magic, as always, belongs to the whole – the combination of the factors that contribute to the overall entity.

Light touch
If you have got this far in the book, you will have had no difficulty in detecting my dislike for organisational boxes. Although I managed through hierarchies throughout my career (because I had not made myself aware of any realistic alternatives), I realise now that there is a significant hidden productivity cost to boxing people in. In a quantum organisation, there is no room for hierarchy. This involves a complete paradigm shift where structures no longer exist to control people but are there instead to enable them. This means that we rid ourselves once and for all of the notion that the guys at the top have all the answers. Because simply put, they do not. The best answers lie close to the front line of the business and those who operate there.

I do not blame existing leaders for this. They still face unrealistic expectations from board members and shareholders to have all information at their fingertips. Not only is this a ridiculous ask in such a fast-moving, knowledge-based economy, but it also creates a ripple effect down into the organisation that ties up so many on information reporting, most of which offers minimal value. In my 15 years or so of coaching corporate leaders, the most consistent complaint I have heard is the amount of time they need to devote to keeping their overlords informed, time which they could spend so much more effectively on leading their people and developing the business. I think we all accept that key indicators are critical to any business analysis and projection, but it really needs to be light touch.

I can anticipate a very mixed response to this idea. I totally acknowledge that if a coach had suggested to me when I was running a corporate organisation that I should get rid of my hierarchy, I would have thought them either crazy or idealistic. When you are dealing with the daily noise of corporate management, it is difficult to imagine such a radically different landscape. But now I know different. There is a better way. It is not an overnight fix, and the transition must be handled with care and skill, but it is doable.

Instead of hierarchy, a quantum structure is built on teams or communities, whichever language you prefer; the terms are interchangeable. Each community is a field of meaning, energetically connected by their shared purpose. The explicit purpose of the community is to be clearly identified. Examples of communities could be business growth, customer service, people development, finance and infrastructure, product and technical, many of which will resemble existing organisational labels. The key difference is not in the labels but in the way they work.

We will each typically be part of a community because we want to be, because it is the natural environment for the expression of our talents and our passions. There are no line managers, only community champions who mentor, support, and enable the team. The champion is the guardian of the team's purpose and vision. Their job is to keep these alive and harness the energy of the community to achieve their goals. Communities will come and go as the need arises. Certain core communities are likely to be a sustainable feature of the structure, while others will be created for a single purpose and be disbanded when the goal has been achieved. We will typically be anchored into one or two teams, but we may find ourselves welcomed to allocate time to other communities to expand our knowledge and learning.

Self-organising

Entirely consistent with the principles just explained is the idea of self-organisation. A quantum organisation does not prescribe exactly what the structure should look like. It devolves that responsibility to its communities. Each community is responsible for defining its own operation and for identifying what links it needs to other parts of the overall organisation. It does not have to link with every other community, but it must hold the door open for any of those communities to engage with its own operation. This is best illustrated by the Figure 3.

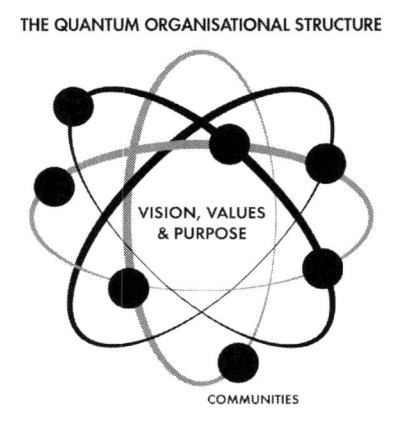

Figure 3: The Quantum Organisational Structure

Here is a visual illustration that is helpful at a conceptual level: seeing the organisation as a planetary system. All systems are energetically linked, with planets orbiting their star and moons orbiting their planets. All are held in orbit by energetic forces. For the solar system this force is gravity: for human beings it's the electromagnetic energy we share and which we feel through resonance and commitment to a common cause. Any organisational structure needs to channel energetic flow, not create barriers to it. The structure is there to support and enable people to perform to their highest possible level. It should feel more like wings than manacles.

In a self-organising structure, responsibility for performance resides with each community; it is not 'the boss's problem'. We will look at this further shortly, but the structural point is that self-organising also means self-regulating. Each community will need to identify its performance expectations and be clear how it will support its members to achieve them. It will also need to be clear how underperformance is dealt with. The role of the overall leadership team will be to ensure there is appropriate consistency across its communities to maintain a sense of fairness, but this is not the same as top-down prescriptive rules and procedures.

Embedded learning
Learning has been a consistent theme throughout this book, a core human need which deserves to be encouraged if we are to achieve our higher potential. The organisational approach to learning needs to be systematic and embedded. It is not a secondary thought, rather a requirement to be nurtured across the total organisation. Remember, the human brain is always learning, both consciously and especially subconsciously. Learning is most effective in real time when we need it. Classroom learning and formalised training can help to motivate and inform people, but there is nothing like learning when you most need it. Learning on the job means that we have the best context in which to store the lesson in memory. It is experiential and more

easily integrates into what we already know. As all trainers appreciate, real learning comes when people take their new skills and knowledge and try to apply them in their work. In this way, our approach to learning needs to be significantly strengthened.

Personal development opportunities need to be regularly scheduled for all community members. This should involve an appropriate blend of coaching, mentoring, and facilitation. This is not necessarily about the more formalised content available through courses but is more about setting time aside to share reflections with colleagues. The organisation needs to up its game in understanding how people have very different learning styles and how development experiences can be better honed to accommodate these needs.

Here's a quick look at some of the key learning styles. A rational style likes clarity, facts, and research, whereas the emotional style responds more to storytelling and bonding. The instinctive style wants quick results and so benefits most from a short-term result-driven experience. The reflective style likes the space to process input and to develop their own ideas.

Sometimes we will choose to take people away from their preferred style and get them instead to learn more from the situations that cause them discomfort. What is important is that we understand the techniques and the people well enough to make the most beneficial match.

Facilitation skills are critical in a quantum organisation and here would be used to support individuals and communities to keep checking in on what they are doing well and where they are struggling. These are reviews without judgement. The emphasis is on lessons learned and honest exchange of perceptions and ideas. Any finger-pointing or undercurrents of blame will destroy the trust and emotional safety that is a prerequisite to their success.

One idea I would suggest is the use of 'trusted triangles',

which are groups of three: two directly sharing feedback and one facilitating. As many people as possible in the business should be trained at a rudimentary level in facilitation techniques. Facilitation is a highly valuable skill that can be used in many situations, both internally and externally, to maintain constructive working relationships. Trusted triangles can be used to support people development and to improve communication and mutual connection between people. When two people in an organisation don't 'get' each other, the relationship normally stagnates or becomes hostile. Traditionally there are winners and losers. The support of a chosen, trusted facilitator can be helpful in moving such a relationship to a more productive footing. This is not mediation, which usually comes after the problem has become entrenched. This is preventative, exploratory, developmental, and creative.

Or it may be a case that two people recognise that they have a particularly important relationship within the community, and, while they get on with each other, they want to use facilitation support to move it on to a more advanced level. It is often very difficult to detach ourselves and genuinely see the challenge from the other person's perspective. The role of the facilitator is to support each party to do exactly that: to help remove the emotional barrier between them and to build a bridge instead.

Such support should be available, encouraged, and normalised, with time formally set aside for such activities. This is about people taking responsibility for their own development and developing the best possible working relationships with their colleagues. We must get away from allowing stigma to become attached to such requests for support. There is an argument that all critical relationships should have a facilitated health check at regular intervals.

Alignment features

Here we look at the features that align the organisation. They sit across both cultural and structural and need to be evident in both.

Collaboration

A quantum organisation has to be collaborative. A sense of equality and openness underpins its daily operations, and it is built on respect for people, talent, and development opportunity. The territorialism and silos of many corporate businesses are the complete antithesis of quantum connection, where doors remain open and people align with their passion, not their organisational section. Cross-community communication is encouraged, and transparency of information and activity are the norm. This acts not only to sustain the sense of the wider organisational purpose but also as a form of resilience.

Even today there are many organisations fighting the battle of having a toxic culture. This remains a remarkably alarming recurrence. Of course, this is a result of many bad practices rather than just one, but genuine collaboration across the organisation can be a strong counterforce to this. My experience tells me that most toxic cultures start at the top, like the fish that rots from the head down, but the problem doesn't end there. In the fight for survival, there will be many other destructive forces operating across the organisation.

Genuine collaboration opens the way for better mutual understanding among people. So often people within organisations are tacitly forced to align with one part of the business and to view the rest with suspicion. This is relatively easily done as we respond most to our everyday experiences, so those in local control have a captive audience. When meaningful collaboration breaks out and colleagues are brought together to work on common goals from a platform of equality, the emotional and instinctive barriers will drop away. What they will see are the same human needs in everyone else: the desire to belong to something that matters and the wish to be acknowledged and to be able to contribute.

An underpinning feature of collaboration is feedback. Well-intended and timely feedback is incredibly useful if handled properly. We all want to know how we are doing, but the familiar tone of corporate feedback has been judgemental. Appraisals and their like have become daunting experiences for so many people when they should be an opportunity for a genuine developmental discussion where neither party feels judged. People have become so wary of offering and giving feedback, and the consequence is an accumulation of missed opportunities and frustrations that usually boil over at the least productive times. The human body could not survive without its own bio-feedback system operating every second of the day; we need positively constructed information from others to help us stay on track.

For me, the ultimate goal in this context is to be able to have completely honest conversations as part of our everyday lives. The current reality is very different. Because of the fears we have built up over the years, we cannot handle such honesty. We are all too emotionally frail, including those who resort to denial or putting on a brave face. Handling genuine feedback demands insight, sensitivity, and skill, but I deeply believe that all organisations should be investing in this capacity for sharing and mutual development. It will take time, but the opportunity for unlocking individual and collective potential is huge. None of us is perfect and we should not beat ourselves up over it, but we can learn to be less imperfect, and this will be accelerated with the help of others.

Diversity

The need for diversity of outlook and action should be self-evident in a quantum organisation. Not only is this the right approach from a moral standpoint, but it is also entirely supported by science. It could reasonably be argued that the original act of diversity was the creation of the male and female of the species. Why was it so? The answer is that we need males and females to widen the gene pool. If we kept on reproducing another version of ourselves based on the same genetic template, we would be confined to a very limited choice of genes and therefore poorly equipped to adapt to the needs of our environment. Diversity gives us the widest possible choice from which to choose the blend that best matches our need to survive and thrive. The same principle applies in the working environment. Why would we limit our choice?

In Chapter 3, we looked at the different behavioural styles associated with the instinctive, emotional, rational, and reflective systems of the brain. If we just slightly change our angle of view here, the same applies to talent. Each of these styles is a form of talent that we can bring to bear in many situations: the gut feeling of the instinctive style, the enthusiasm of the emotional style, the diligence of the rational style, and the vision of the reflective style. Each of these is just one example of the amazing talent that our brains are capable of. The term 'neurodiversity' became popular in the 1990s to reflect just this. Talent has no predefined sex or ethnicity. And so, a quantum organisation nurtures diversity of thinking, recognising the limitations of the predictable and aspiring to the potential of the possible.

We need to be very conscious of our own biases and prejudices. The term 'unconscious bias' has become a buzzword in the industry, regrettably often approached with only superficial intent. We are all unconsciously biased; there is no other option. Our brains cannot process all information that is available to us in an unbiased way. We subconsciously learn from birth what to pay attention to and what is important, what to be attracted to and what to avoid. The challenge of self-discovery is understanding what our biases are, where they came from, and what we can do to release ourselves from them.

A quick look back at the history of the scientific study of the brain illustrates how deeply held some of these prejudices can be. Throughout the 19th and even early 20th centuries a great deal of research was undertaken to discover the secret of what makes the male brain smarter than the female.

Note this was not about finding out if they were or not; it was a standard assumption within the scientific community that this was the case. But could they prove why? Despite millions of dollars being set aside for the research (of course sponsored by male-dominated institutions), no credible evidence has ever been produced to support the assumption of male neural supremacy. There was some short-lived excitement at one point when one body of research claimed to establish that this was down to male brains being larger. This was then rapidly discredited as the size difference was found to be entirely proportionate to the respective body sizes of men and women. (We also know now that the connectivity of the brain is a much more reliable indicator of proficiency than size.)

I would like to feel that the days of such blatant prejudice have passed, at least in most organisations, but the subtle prejudices of modern living still prevail. At a time when the Black Lives Matter movement has been so effectively mobilised, we must all reflect on institutionalised prejudices and what actions we can take to implement change. Similarly, the courage of those who created the MeToo movement to stand up to sexual harassment and the abuse of power. Such abhorrent prejudices have no place in the quantum world.

Ownership and accountability
I put these terms together because I see them as two sides of the same coin. I have been involved in several organisational discussions where the meanings of each appears to become blurred, so I will offer my own distinction. For me, in this context, ownership is a personal sense of responsibility, something we feel belongs to us. Accountability is the external expression of this responsibility, such that we are held to account by others.

This feature made my top list because it is such an important consideration for anyone running an organisation. I am aware that many business leaders' instinctive fear about the quantum concept is that it appears to leave the need for performance to chance. It seems to be all about flow and collaboration and other nice things. I can relate to this: when I lead organisations, I was very keen to know who was accountable and who appeared to demonstrate that essential sense of ownership. So, I would like to offer some reassurance.

The need to perform, and to be acknowledged for performing, is built into all of us. We are designed to seek out opportunities to thrive as part of our evolutionary momentum. Affirmative feedback gives us the dopamine hits we need along the way to keep us going. This is, of course, true of many individuals in today's organisations, but the key difference is whether or not it is a collective phenomenon. In the traditional organisation, there are individuals who show great resilience and ambition, but these individuals are largely selfish and have to be because they are set up to compete against each other. In the quantum organisation, the same resilience and ambition is encouraged, but it is for the good of the community, a far better place for emotional safety and growth.

High-performing teams are self-regulating. They want to

perform at their best, both individually and collectively. They have a clear shared vision, trusted relationships, and purposeful machinery, by which I mean the mechanisms for getting the job done. (See Figure 4 for the podium model which illustrates this.)

THE PODIUM

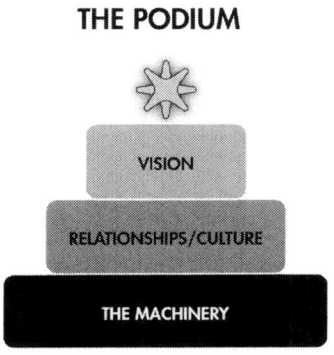

Figure 4: The Podium Model

Professional sports teams are a great example of this. Their shared vision is clear: winning the competition they are part of. Being able to look around the dressing room and see players who trust each other and want to be part of the team is critical. A natural order will evolve which is based on trust and friendship, not hierarchy. The more senior players will take a slightly different influencing role in the team to that of the young rookies. There is ample room for people to excel. Top performance is recognised by teammates who share in the glory, and less successful days are met with support and encouragement. When the team is in play, everyone needs to instinctively know their job and the job of their teammates, and to trust them to give their best.

There is no hiding away from the results in sport. No provisions or accruals to smooth out the performance perception for the external stakeholder. This means that honesty among teammates is critical. I have observed on many occasions the sometimes brutal honesty that colleagues in play will offer their teammates. This is acceptable when there is trust and when the team can return to a more supportive dialogue after the game. Criticism is okay if it is clearly for the good of the team and everyone knows it, as long as it doesn't become intentionally destructive or over-personalised.

The same applies to artistic productions. The creation of a theatre play is a massive team effort where the cast and crew are on a journey together, where emotions are entwined, and the ups and downs of the artistic process are lived with real intensity. Success relies on everyone taking responsibility for their role and their obligations to the whole team.

Likewise, high-performance organisational teams will find their own way of sustaining performance standards. If the team accepts underperformance, it will become unsustainable. Individuals who struggle to keep up need to be supported. On the other hand, over time the team cannot be sacrificed for one individual, and this will mean in some cases that it is time for the individual to find a better match for their purpose and talents. Ideally this could be addressed through another community in the organisation; if not, then it will need to be in another organisation altogether. Such exits should be handled with grace and humility. All teams are temporary and the process of joining and leaving should be entirely natural and unthreatening.

Empowerment is one of the central thrusts of any quantum organisation, and herein lies the valuable upside for the wary CEO. Such empowerment, nurtured by the whole of the organisation, and not just by a formalised and imposed authority system, creates the opportunity for organisational leaders to have freedom to lead, a point we shall develop in the next section.

Quantum leadership

The quantum organisation gives leaders the chance to lead. Of course, they need a level of 'machinery' (refer to the podium model, see Figure 4) in place to support the organisation. All operations need some rules, but they are not to be enslaved by them. Reporting requirements are reduced as more can be achieved through open and honest dialogue. Leaders who rely on formal authority do not belong. The quantum leader's role is about mentoring, nurturing, coaching, facilitating, supporting values, maintaining purpose and belief, and, above all, inspiring.

Most leaders I have worked with would welcome the opportunity to focus on these aspects and to redefine their working relationship with their colleagues to one where they play much more of a strategic and guiding role. This does not mean that they cannot be hands-on. Hands-on should be about maintaining close relationships with key colleagues and being there for them supportively. It is not an excuse to micromanage and swamp them in data reporting requirements.

The role of the leadership team is crucial in maintaining the energy and resonance of the whole community, the gravitational pull of the star in the solar system that keeps everything aligned. They are the people who are in the position to observe patterns right across the business, to see the things that others do not see. They are there to energise, articulate, and nurture the vision of the business and to ensure it is felt by every community within it. Their visionary role is futuristic. Here we are not talking about forecasting and planning, but a deeper intuitive ability to conceive future possibilities.

Quantum leaders need the imagination to see into the future beyond what is immediately in front of them. This touches on an intriguing scientific point. Now is not the place or time (pardon the pun) to get into a complex discussion about the relationship between space and time, but suffice it to say that research is starting to show us how some people genuinely seem to be better neurologically equipped to look into the future. The universe works like an infinite matrix of possibility: energetic intersections that stretch out before us, to be encountered or by-passed as we move forward. Research is showing us that some people are better equipped to sense and see these intersections, and therefore have a stronger feel for future patterns and opportunities. It shows successful entrepreneurs to be among the best in exercising this capacity.

This is underpinned by our rapidly growing knowledge of the brain: in this particular case, it involves the integrated capacities of the cortical lobes and the pineal gland, which works like our Wi-Fi system linking us to outside energetic forces. We will return to this subject in the next chapter.

Like all teams, membership of the leadership team should be flexible. There are likely to be those who stay around longer for purposes of sustainability and continuity, but no role should become institutionalised. Then there are those who will move in and out of the team as their talent and purpose evolves. Membership of the leadership team is about having the right talent available at the right time to lead and support the total enterprise. Again, entry into and out of the team should be a natural process of adaptation and diversity.

Despite the size of the challenge for leaders, I am very encouraged by those who are starting to show interest in the quantum ideal. It may be unfamiliar, but many good leaders get this at an intuitive level. They are already doing some of this stuff without using the same language. Their difficulty is in piecing all of this together in a way that seems coherent and communicable. They tacitly understand that there is a need for a paradigm shift in organisations. Incremental initiatives have been done to death and attempts at significant change have been slow and marginal. Transformation will need to be skilfully directed and led energetically by those who have the belief to inspire.

My advice is based on the universe itself: believe in what you are doing, get an outline plan in place that gives all people in your organisation a real opportunity to understand and engage, and then get on with it, knowing experience will have many lessons waiting for you, and how important it is to stay open and adaptable to them. Above all, do not accept failure: there is always another way.

Quantum in action: a case study

Now let's turn to The Happiness Index as a case study to understand what some of the challenges are on the ground when transforming to a quantum organisation. Here we have a summary of a discussion held between two members of The Happiness Index's leadership team (Jackie Dyal, Human Experience Director, and Gemma Shambler, Head of People) and their honest reflections on what they have experienced so far and what may lie ahead. They are representative of the types of challenges an organisation will face, regardless of scale.

The full discussion can be heard on the podcast, found here: https://open.spotify.com/show/7h6wCWnjsgOIZSR6h0DKbq.

What is the quantum way?
The big thing to remember is that the quantum way isn't a single thing, there's not a definitive answer as to what it is – it's going to look completely different for every organisation. But, at its heart, the quantum way is a way of working which will help benefit a company's people, and their bottom line.

At The Happiness index, the quantum way is about defining our own journey and finding a way for us to collaborate and evolve in a non-hierarchical way, alongside each other. This allows us to thrive, because it allows our culture to thrive. It's about allowing individuals to thrive within the universe of quantum energy.

Why the quantum way?
We want to remove the command and control approach that is so detrimental to the workplace. The question then is, with what do we replace it? We believe that quantum principles, which is how the universe works, and neuroscience, which is how we work, are two sides of the same coin. So, we can use what we've learned about energy and directing it and applying it in the quantum way.

Instinctively when we heard about this way of doing things, it really resonated with what we wanted to create and promote in terms of building a thriving culture for people. Fundamentally, the quantum way of running an organisation means that, at its core, it's a very people-centric place to be, which struck a chord with us. It means that the energy that people bring to the organisation isn't hampered or boxed off, but instead is directed in such a way that brings the best out in people and encourages great connections across the business.

What is the Quantum 9?
The Quantum 9 (Q9) are behaviours and principals that we try to live by when we're implementing a quantum way of working within an organisation. They're the building blocks which allow a quantum way of working to flourish. The Q9 are:

- ***Light touch****: Everything is very flexible and agile*
- ***Self-organising****: This helps with agility and the ways teams are made up*
- ***Embedded learning****: Opportunities for growth and development at all stages*
- ***Trust****: This is the massive one that everything is built upon – you can't implement this way of working if you're not willing to be open, honest, and transparent with each other*
- ***Self-aware****: Giving and receiving feedback*
- ***Collaborative****: Making sure we're working with one another well*
- ***Ownership and accountability****: Helping the organization to be-*

come very outcome-focused and giving people the ability to take responsibility and own their decision-making process
* ***Diverse:*** *In all its forms*
* ***Personal growth and meaning****: If you're happy with where you work and you're happy at work, you're going to be able to perform at your best*

What have been the biggest challenges?
Implementing the quantum way hasn't been perfect. Often when we think with our rational brain, we will think that there's a roadmap for rolling out the quantum way, and we'll know when we have achieved our goals. One of the biggest challenges we found at The Happiness Index is that we can't see in detail where we'll get to and how we'll get there; we just know that it will probably be a feeling.

This is quite a different and unusual way of working in a corporate or work environment. Usually emotions and feelings are left out of our places of work, so we're breaking the mould here. So, when we think about what quantum is going to look like, we instead need to think about what it's going to feel like. One of the biggest challenges is that there's not an end point to this implementation – it's a journey.

Neuroscience is about understanding emotion as well as the rational. We have to look at individuals and get their feedback. We have to get them to piece together, and work through all the different brain types and bring them together. For example, we're trying to work in a more agile way and be less hierarchical. However, one of the challenges in this is that the people that need to be empowered may feel like they don't have the support that they would usually have in a normal hierarchy.

So, we have to think about putting frameworks in place to help people get there. For example, following feedback, we have been putting in place mentoring instead of management. This is something we've learned from feedback, which is going to be key as we go through this journey.

One of the things we noticed when implementing the quantum way is that because there isn't a prescribed end goal, it can at times be hard to know how to drive it further and take the principles that we believe in and implement them better. And we're still on this journey, and there are still elements that we're continuing to implement, for example bringing in structures for those who do need more of a framework.

Everyone is going to be going through a different change as the quantum way is implemented. A key thing to remember is that there's a change curve process that we all go through (see Figure 6 in Chapter 7) which can be expressed in neuroscience terms, and so some people will get to a sense of meaning and purpose at different times as we go through the process.

Everyone who takes on the model is going to feel it in a different way. One of the ways this could be felt is fear because you're moving from a model that you know – and may feel works – and moving to something new and unknown.

Is it worth it?
This comes back to the feeling. It certainly feels like there's a change and a shift. Before we embarked on this journey, we felt that The Happiness Index was an open and transparent place to work with high levels of trust, all of the elements that we've discussed being principles of the quantum way. But now since implementing more of the quantum way, and even since the beginning of 2021, this has gone up another whole level. We've found that the amount people are sharing, and bringing into work, has changed entirely. Even though we've always prided ourselves on allowing emotions in the workplace, people were still holding back a bit, but now we're seeing more and more of people sharing.

What is a quantum board?
One of the fundamental pillars of quantum is diversity. One of the facts of The Happiness Index is that the three founders who formed

the board are white men (arguably also middle-aged, but we can pick that up separately!). We're now working to make our board quantum.

Fundamentally, the purpose of the quantum board is to make sure that they, as a group, are providing the conditions for The Happiness Index to have a thriving culture and be able to deliver a sustainable level of growth. They also need to be adhering to the Q9 as much as possible. As with most boards, they are responsible for setting the strategic vision for the company and giving clarity and enablement to our people.

Giving Clarity, Empowerment and Equal Opportunities to All

The Quantum Board will be guided by the Q9 Principles:

THE QUANTUM 9	CULTURE Ownership & accountability Growth & meaning Trust	ALLIGNMENT Self-aware Diverse Collaborative	APPROACH Light-touch Embedded learning Self-organising

SEAT	TYPE	PURPOSE	PERSON	VP
Employee Happiness & Culture	Fixed term (2 years)			
Employee	Fixed term (2 years)			
Customer	Rotates quarterly			
Tech	Fixed term (2 years)			
Growth	Rotates quarterly			
Finance	Fixed term (2 years)			
Chair	Fixed term (2 years)			
Founder	Rotates quarterly			
Owners/Investors	Rotates quarterly			

Figure 5: The Quantum Board

Our initial motivation for implementing the quantum board was to make the group as diverse as possible in order that they can be as successful as possible. This is because we know that diverse teams perform better. In order to do this, we've introduced a few different roles within the board, with the aim of introducing different voices, opinions, thoughts, and experiences to support the conversations that take place.

On our board we have our founders, a chair, and a representative for our investors, as well as a financial representative – all the typical positions you would expect to see on a board. What we've also now included is a representative for our own employees' happiness and culture. We also have a different employee join the board every quarter. This is an individual who doesn't usually have board responsibilities or reporting responsibilities to the board. They are simply required to come along, listen to the conversation, and contribute. This helps with their own development as well; it allows them to listen to the kinds of conversations that take place, talk directly to the board and investors, and so on. We've also had a customer sit on the board to support the conversations.

At The Happiness Index, we're always looking for ways to make our board more representative, so we're looking at adding someone to represent our tech team, which is important to us as a tech company. We're also going to include someone from our growth team. A number of these roles will rotate quarter on quarter. We're hoping to rotate this through the different members of the teams, so no one person is appointed to the position and it's theirs to keep. Instead, we'll bring in different members of the team, different voices, and different opinions, depending on the subject matter at the time.

What advice do you have for others trying to implement the quantum way?
Gemma's advice is to be brave and stick with it. There may be more than one point where you find yourself asking yourself if it's working or whether it's worth it. While we've not quite reached a point where it feels as though everyone hates it, you definitely do get some doubts sneaking in about whether it's the right thing to do, or whether you've gone about it in the right way. But stick with it!

Keep challenging yourself, keep asking questions of yourself and your team about why you're doing things or how you're doing them.

Keep applying the principles, say: 'No, that's not the quantum way; we need to flip it on its head and do things differently.' Don't be afraid to keep challenging the norms and the traditional ways of doing things and thinking.

Jackie recommends starting with an education piece, as this will help to cement an understanding of the journey with the whole team. At The Happiness Index, we were lucky to have an in-person workshop, which we made as fun as we could, we really were able to embed the Q9 and the behaviours into the team and help them understand the overall direction. This helps get everyone aligned.

One thing we didn't do, but may have been useful, is map out how far along the journey we already were. This could have helped manage expectations among the team on how far along the journey we already were. Jackie would advise people to assess where they are in terms of the quantum way already and share this with their team. Another expectation to manage is to underscore the idea that there isn't a defined end goal, that there isn't a North Star. It's important that everyone understands that there isn't a deliverable or a plan, but instead the journey will be based on constant evolution and behavioural change. Don't give yourself too much pressure to create change officially; let it evolve through behaviours.

A quantum approach requires significant rethinking for all involved and especially for leaders. The things we have become accustomed to – hierarchy, passing the problem up to the boss, paralysis by analysis, and approval hurdles to be jumped at every turn – are dismantled and other mechanisms put in place, allowing talent and passion to flow into the organisational purpose. If we strip away the controls of most organisations, it is perfectly understandable that our first reaction may be one of fear and confusion, but when we ground ourselves in our own refreshed energy and feel uplifted by the energy of those around us, we will feel released to pursue the opportunity to reach our higher potential both individually and as a collective force.

Having explored an internal organisational perspective, we can now look externally at what has been happening globally during the pandemic. Although a very difficult experience for so many people, the Covid-19 pandemic has given us an opportunity to capture data about the human experience which is unique. Matt Phelan will take us through some of these insights.

CHAPTER 7. HAPPINESS AND ENGAGEMENT DATA

Hey there, Matt Phelan author of *Freedom To Be Happy* returning from the preface of this book (in case you skipped that bit) as a cameo in this chapter.

All of this exciting stuff Clive has discussed so far is hopefully appealing to your instinctive brain, but here is some data and evidence from the frontline to appeal to your rational brain.

At The Happiness Index, an employee happiness platform working across 90+ countries, we have been using ideas from *The Quantum Way* for the last two years. We have been collecting data on how people feel from millions of employees every single day in real time. Here I have collected some of the stories from our journey, stories and data ranging from very small companies to some of the biggest organisations in the world.

What is it like to actually implement the quantum way? And can it work?

It is not easy to implement the quantum way. Anything worth doing takes time, effort, and commitment. There have been times over the last two years when we've been working in a quantum way, and I have wondered if it is all worth it. However, what we are getting from it is huge, and I can feel that we are unlocking resources of untapped potential within ourselves at The Happiness Index. Our own neuroscience change curve (an update on the Kubler-Ross change curve) probably best describes my feelings on moving to a quantum way of working (see Figure 6).

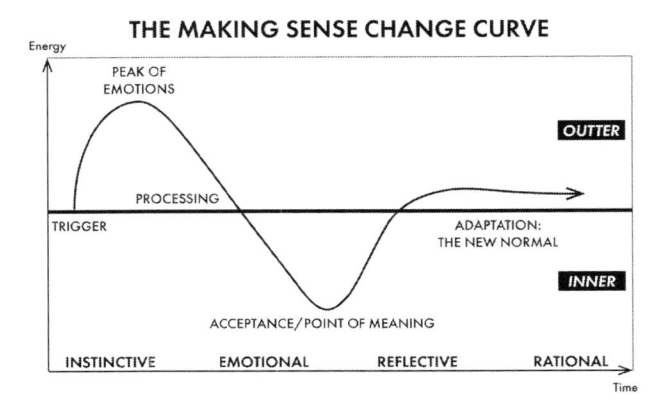

THE MAKING SENSE CHANGE CURVE

[Figure: A curve showing Energy over Time. Labels include PEAK OF EMOTIONS, PROCESSING, TRIGGER, ADAPTATION: THE NEW NORMAL, ACCEPTANCE/POINT OF MEANING, OUTTER, INNER. X-axis phases: INSTINCTIVE, EMOTIONAL, REFLECTIVE, RATIONAL.]

Figure 6: The Making Sense Change Curve

It is not happiness vs engagement but happiness + engagement

One of the outputs of working in a quantum way is that it has helped us produce the happiness and engagement model. For years we couldn't work out why so many people contacted The Happiness Index saying they had sky-high employee engagement scores but their staff were unhappy, people were leaving in droves, and they had record levels of staff reporting bullying and mental health issues. They were spending hundreds of thousands of pounds on employee engagement technology, and it wasn't helping. It was giving them the right scores and flashing lights, but it wasn't helping their people.

In the world of sport, England U17's football coach Justin Cochrane calls this "comfort coaching", aka doing things that make you feel comfortable and in control but do not necessarily help the team. More from Justin here if you want to avoid what he describes as "flat pack coaching": tinyurl.com/flatpackcoaching.

Quantum insights, neuroscience, and our real-time data

helped us realise it is not about employee engagement vs employee happiness but employee happiness + employee engagement.

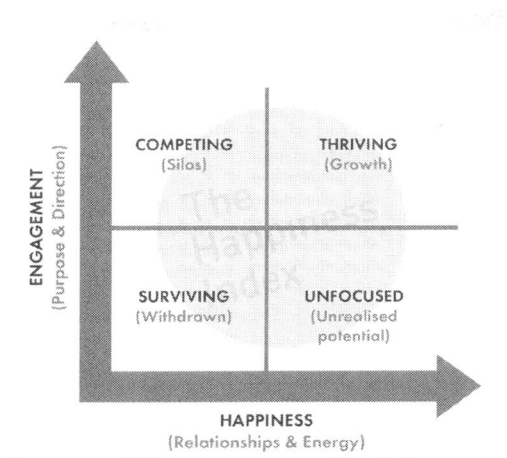

Figure 7: Happiness and Engagement Model

Many people credit an academic named William Kahn with inventing employee engagement, so I caught up with him recently on the Happiness and Humans podcast to put the record straight. Following Kahn's 1990 paper "Psychological conditions of personal engagement and disengagement at work", an entire industry was created called employee engagement. Kahn discussed how he believes his version of personal engagement is violently different to employee engagement and has been co-opted by the corporate world.

I fully support Kahn's view, and, in our world, we believe it is about reconnecting the heart and brain of business. Allowing emotions back into work is not unprofessional and is important if we want human beings to thrive at work.

Figure 8: Reconnecting the Heart and the Brain

So, why do many companies not do this? I blame this statement: "If you can measure it, you can manage it." In the 1990s, people couldn't measure feelings and emotions because the technology at the time couldn't handle it. People were also actively discouraged from sharing their emotions at work.

We measure emotions at The Happiness Index not to manage them but to understand them. The more you understand the emotional state of your business, the better your decision-making. In my opinion, the better your decision-making the better your leadership.

Today's emotions are tomorrow's performance. Step one to understanding this is visualising your culture.

Measuring emotions and what we see
Surviving
In organisations that are just surviving, we see low levels of happiness and engagement. They are in survival mode and tend to be very inward looking. Survival is an important stage for an organisation and is part of the evolutionary process. For example, many businesses in the retail and travel sector have had to spend over 12 months in survival mode due to the Covid-19 crisis. To be in this stage for prolonged periods of time is not sustainable, but it can be important for survival.

Unfocused
Organisations with high levels of happiness but low engage-

ment tend to be unfocused. People are happy despite the business. They invest their energy in local relationships or outside interests but not in the business itself. This results in a lack of direction and alignment.

We sometimes see this in charities and not for profits organisations where there is a lot of positive intent but a lack of strategic direction.

Competing
Competitive cultures are often mistaken for thriving cultures. Insight from these types of organisations is one reason we dropped the phrase 'high-performance culture' form our internal language at The Happiness Index.

Competitive cultures have low happiness and high engagement. They have personal agendas rather than shared ambitions, and a lack of strong relationships means the organisation works in silos. Employees end up competing in high-pressure environments that we often see associated with job roles that include sales and recruitment. These companies succeed not because of the competitive nature but in spite of it. And at what cost?

The best salespeople I know are great collaborators and buck the stereotype of a competitive salesperson working as a lone wolf.

Thriving
Thriving cultures have a good blend of happiness and engagement. Organisations that are thriving have a clear sense of direction and purpose and are energised by strong relationships. Organisations with a high percentage of employees in the thriving category will tend to have high levels of both organisational growth and individual growth. These don't need to be mutually exclusive. Business is not a zero-sum game, and, in the right en-

vironment with a thriving culture, both the individual and the organisation can win together.

As Obi-Wan Kenobi taught us in *Star Wars*, it is about balancing the force. A happiness + engagement mindset has helped us build technology that balances employee happiness and employee engagement at scale.

If you can balance happiness and engagement in your business, you have a chance of creating a thriving organisation. Happiness and engagement does not need to compete – after all, only Sith Lords believe in absolutes (I promise that is the last *Star Wars* reference I'll be making!).

Happiness in a pandemic – 12 months of data

It is impossible to write a book in 2020/21 without mentioning the global pandemic and how it has impacted how we feel. As Covid-19 impacted different countries at different times, for this section, I am primarily using UK data to tell the story on how employees are feeling and so give you a linear timeline.

Months 0–3

Unsurprisingly, during the first lockdown, our happiness data revealed that happiness took a sharp downturn in the first three months of the pandemic. Prior to Covid-19, we'd seen average scores of 7.3 which fell sharpy to 5.8 in the first month before reaching a 'new normal' at 6.1. What was clear was that people felt the need to share more, with participation increasing to 69%, almost double what we'd been seeing before, and the number of words left in comments increasing from an average of 10 to an average of 37. We call this an emotional deficit and something organisations can either ignore or help with.

Generally, while people were enjoying increased time with their families, and having more time to exercise and enjoy hobbies, the uncertainty of this period was what was worrying people. Not only this, but many felt that they had to work harder and for longer hours while working from home. We could also see that the good weather we enjoyed during this early part of the pandemic had a great impact on people's moods!

Months 4–6
What we found as we neared the six-month period of the lockdown was that morale was getting lower as time went on. The new normal didn't stay as normal as the pandemic dragged on, and happiness scores fell to a low of 5.2 at the end of September, with an average score for this period of 5.8.

It's clear from people's comments that the continued uncertainty was having an effect on people's stress levels and mental health. Not only were people still sharing at least twice as many words as they were pre-Covid, the comments showed that many people felt the need to work harder than ever and pull their weight. The impact of the lack of downtime and people not taking adequate days off over the summer period was also clear.

Months 7–12
Despite the fact that the festive season looked very different for most in 2020, we still saw an uptick in happiness over this period, with an average happiness of 6.2, similar to the 'new normal' established in the first lockdown. However, this dropped back down to 6.0 after the festive fun and breaks finished.

Comments during this time showed that the lack of light, poor weather, and seemingly unending lockdowns were having a real impact on people's mental health. This was particularly the case for those with school-aged children who were struggling to balance home schooling and work. Nevertheless, there were glimpses of light at the end of the tunnel. As we reach the 12-month mark in this pandemic, strong vaccine rollout is driving more confidence in employees.

Covid-19, one year on
My prediction is that as we come out of the first 12 months of the pandemic, we will see one of the biggest ever migrations of employees.

Employees that have been looked after are more likely to remain loyal and thankful for the way they have been treated. Companies who have not listened to their concerns will see employees resign in record numbers.

We will move into an era where employee brand and transparency will be even more important than they are today.

Emotional deficit

One other point on the emotional deficit we are seeing in the data is to understand the difference between fatigue and depletion. To better explain the difference, here are some thoughts from leadership and transformation expert David Lapin:

Fatigue is when a muscle or the mind has been used almost to the point of failure and it requires rest to recuperate ... Contrastingly, depletion is when our inner resources of energy have been drained – and this may have no connection to exertion ... Consider an empty glass of water. You can rest it for as long as you will, but rest won't replenish its contents ... When we are depleted, we need restoration rather than rest or relaxation ... Many of us are currently suffering more from depletion than from fatigue, but few recognise this. Some took summer leave and came back as exhausted as before, because rest and relaxation alone don't restore depleted inner resources.

We know from our data that relationships are the number one driver of global employee happiness at work, and it will be important for companies to consider how we support strong relationships to help employees restore emotional reserves.

GDP to employee happiness conversion rate across the globe

As we discussed in *Freedom To Be Happy*, money does make you happier to a certain extent, but in a developed country the impact of money on happiness lessens. If you live in a developing country, more money can give you access to basic needs like clean water that can drastically increase your health and happiness. However, in developed economies, more money does not have a similar impact. This is not to say some of these economies do not have major inequality issues, but the impact of money on happiness is less.

For this book I have reviewed the top 11 economies (I have no idea why I did not make it the top 10, but there you go). See Figure 9.

	COUNTRY	GDP	AV EMPLOYEE HAPPINESS INDEX FROM THE LAST 12 MONTHS
1	USA	19,485,394,000,000	7.4
2	CHINA	12,237,700,479,375	8.3
3	JAPAN	4,872,415,104,315	8.9
4	GERMANY	3,693,204,332,230	7.2
5	INDIA	2,650,725,335,364	7.2
6	UK	2,637,866,340,434	7.2
7	FRANCE	2,582,501,307,216	6.9
8	BRAZIL	2,052,594,877,013	7.2
9	ITALY	1,943,1835,376,342	7.0
10	CANADA	1,647,120,175,449	7.2
11	RUSSIA	1,578,417,211,937	7.8

Figure 9: The Top 11 Economies' Average Empolyee Happiness

As you can see, Japan comes out on top, removing China from the pre-pandemic league table of employee happiness. I do wonder if there is something we can learn from Japanese culture about resilience. It may also be that China has had a harder time with Covid-19. While it is too early to confirm the reasons for these changes, we should be considering and discussing them.

We can see from this table that GDP and happiness in the workplace aren't directly correlated, particularly in the case of Russia, which has the lowest GDP of the countries represented in our data, but the third-highest average employee happiness. By contrast, we can see that despite the UK being firmly in the middle of the table when it comes to GDP, it has one of the lowest average employee happiness scores, and France has the lowest GDP to employee happiness conversion.

Correlation does not = causation. But a correlation is a useful starting point to ask questions about the relationships and associations between money and happiness.

Global drivers of happiness and how they change from country to country

One of the reasons we measure emotions like happiness is because happiness is a global emotion that all human beings experience and can relate to. Engagement is a useful business metric, but it is also a very Western-focused term.

To test this concept, try meeting a hunter-gatherer tribe in the Amazon rainforest and attempt to explain to them the concept of employee engagement. My bet is you will have a much better chance of explaining happiness.

As we collect more and more data in almost 100 countries, it has been super interesting to see how happiness as an experience is very similar, but what drives it changes. We don't have time to go into all the differences, but in order to gain an understanding of how drivers of happiness change, let's take three countries that while historically linked are also very different: the UK, the US, and Canada.

The UK and the US share many similarities, such as both

ranking 'relationships' most highly as a key driver of happiness for employees. However, there are also significant differences: the UK employee ranks 'clarity' of role as a low driver of happiness, whereas in the US 'clarity' of role is a key driver of employee happiness.

You may expect to see similarities between drivers of happiness with close neighbours, for example Canada and the US. In the US 'acknowledgement' is the lowest driver of employee happiness, while in Canada it was the most important factor. So, for an American employee clarity is a real driver of employee happiness, but for a Canadian worker, acknowledgement is key.

Although on the surface these factors might seem like small differences, they highlight the need for cultural understanding in your business and HR strategy.

Summary

Measuring happiness is a very quantum thing to do, but I recommend you don't view it like an old-fashioned sales report where high sales are good and low sales are bad.

High happiness does not always mean good, and low happiness does not necessarily mean bad. They are just information points to better understand the health of your organisation so you can make better decisions.

We have clients whose workforces range from 100 employees to 400,000 employees, and when they start reporting happiness in a board report, I often ask them to see it more like a weather report. Rain in a weather forecast is not a good or a bad thing. A forecast of rain for a farmer can help them plan irrigation levels, but for someone like me who lives in London this information can ensure I pack my coat with the hood. Context is key. Remember, working in a quantum way is more like an ecosystem and less like a factory shop floor.

We created the follwing visual (Figure 10) with the help of

our amazing graphic designer Joe Wedgwood to help people see this change in mindset.

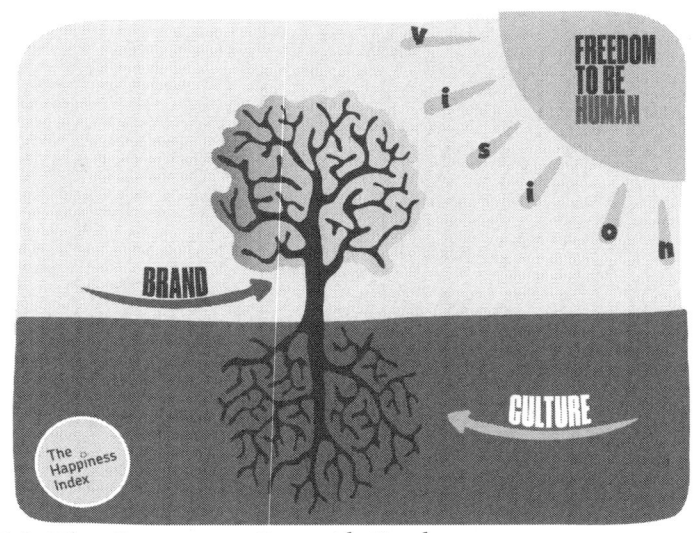

Figure 10: The Quantum Growth Cycle

In simple terms, your vision should energise the company and your culture should be the roots that your entire company grows from. Brand and culture are part of the same ecosystem. In a world of websites like Glassdoor, TripAdvisor, and Trustpilot, brands need to flow from their culture, or they will simply be exposed as frauds. A quantum mindset acknowledges that everything is linked.

If you get the quantum way right, it can be one of the most energising feelings. This is because we believe and feel that the quantum way is taking us closer to our vision of what we call 'freedom to be human'.

If you want to continue the conversation with me, please join The Happiness and Humans Community here:tinyurl.com/joinhappyhumans.

And with that, I will now hand you back to Clive.

CHAPTER 8. THE SCIENCE OF BELIEF

Having spent the last chapter interpreting empirical data, we now return to the non-empirical, to the realm of belief. We cannot measure it in a rational way, but even the smallest amount of reflection will highlight how significant a role it plays in our lives. Chapter 4 showed us that not only do we typically believe what we see, but we also only see what we believe. Our beliefs frame the way we see the world, and this perspective acts as a catalyst for our subconscious responses. The deepest self-understanding can only come when we are able to step back from our own beliefs and explore their influence on us.

Can science help us to understand beliefs, how they are formed and how they influence us? It can certainly throw some light on this, but such a subject area is essentially subjective so we cannot at this stage rely on science alone. Such a path of enquiry also needs an intuitive inner exploration and a wider look at the influences that shaped our beliefs. Earlier in the book, I referred to the influence of parents and early caregivers in shaping young brains: now we shall look at the influence of spiritual and religious beliefs.

Such a topic is widely disregarded in Western business leadership literature, but I believe this is a mistake. The influence of such beliefs can often be very subtle, passed on through generations, but that should not blind us to its power. We are not here trying to resolve the validity of such beliefs, but we are trying to understand the role they play in our lives. Most of all, it should never be a 'no-go' area for discussion. How can we possibly hope to build successful and happy integrated communities and organisations if we do not understand other people's view of the world, particularly across diverse global cultures? Ignorance breeds misunderstanding, fear, and judgement, which all separate us. We need to strive to understand each other without prejudice and with complete respect for the causes that we all believe in.

We need greater awareness of all internal and external forces that inform our view of the world, and this must include our belief systems. And here I am not talking just about our habitual beliefs we carry around day to day, but more our ultimate beliefs in life and its purpose. This is where recent developments in scientific thinking can be very helpful. Throughout the more recent years of my study I have been struck by the convergence of thinking between the science of the known and belief in the unknown. The language may be different, but the underlying concepts seem very similar. This is explored further below.

I am privileged to be doing some work with Caplor Horizons and Paper Boat, UK-based charitable organisations working with partnerships across the world supporting people facing huge social and economic disadvantage.

For example, the Dalits, the lowest caste in Hinduism, formerly known as 'the untouchables', are told from birth that they are unworthy, and that they deserve only the lowest social status. Such indoctrination traps them in disadvantage, a convenient protection of the status quo by those who benefit from their persecution. For many people, their beliefs can be positive and lift them spiritually from their daily burdens, but they can also hold them back to accept their allotted place in society.

Whilst I can loosely deal with both spirituality and religion here in the same way, I should offer some distinction. For me, religion appears to entail a belief in a divine power and requires adherence to a set of rituals and practices. Spirituality also involves a belief in some meaningful force which is bigger than us but with much less emphasis on ritual. Both require an inner journey of reflection and meaning. There is clearly overlap between the two, and, in my view, both are to be respected.

I am not trying to come to any existential conclusions; rather to understand the impact our beliefs have on our lives. In trying to understand people there are many layers to strip away, from the behaviours we express, the internal dynamics of our subjective experiences and the role of memory. If we kept stripping away, what would we find? If we were somehow able to define all the influences on us, what are these forces acting on? What is our *source*? What are we at the most pure, uninfluenced level? We touched on some of this in the self-discovery section, but now we will look again from a more universal perspective.

Convergent thinking and consciousness

Convergent thinking starts with an acknowledgement of a greater universal force which significantly impacts our lives. While spirituality typically acknowledges the presence of some form of divine being, quantum physicists refer to the infinite energy of the universe. We hear of omnipotent gods and of quantum energy being all that there is and ever has been. According to quantum physicists, the quantum vacuum existed in the split second after the Big Bang and before the infinite potential of the universe materialised in physical form. Quantum energy is the essence of all being. In religious terms, God may well be referred to in similar terms.

The difficult question for scientists is: what existed before the Big Bang; what created quantum energy? In religion the parallel question might be: who created God? We simply cannot conceive of such answers. Ultimately, whether scientist or spiritualist, we rely on belief. In this sense, belief serves as an anchor to ground our internal processes for making sense of our lives. Without belief we face the infinite with no shape or form. If the infinite is inconceivable, what is within our conceptual grasp?

Arguably, the closest we have come so far to scientifically understanding these big questions is the subject of human consciousness. Here we link back to the discussion about mind and brain. We have obvious evidence of a physical brain and are learning more and more about the way it works. But we cannot understand it completely without understanding its unobservable energy systems. We know that the mind is an energetic phenomenon, but we are only just uncovering how it works. We know, for instance, that internal visual imagery is critical to our interaction with our environment and to our own subjective experiences, but it cannot be pinned down to any one part or parts of the physical brain. Neuroscientists refer to the retinoid stage as the internal theatre of the mind, a complex energetic matrix of holograms and energetic planes.

Consciousness is one such energetic pattern. Having said that, we need to distinguish between consciousness and awareness. While there is indeed overlap, awareness is a term generally used when we are cognisant of something in the moment at a conscious level. There are at least three levels of awareness: starting with a sensory sensation in the body, then moving to conscious awareness of this sensation, and then to our awareness of 'self' (see Figure 11).

AWARENESS LEVELS

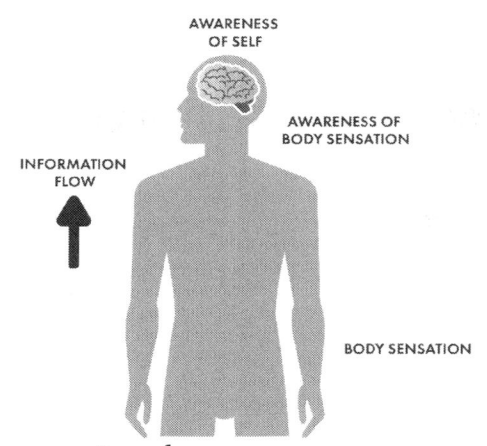

Figure 11: Awareness Levels

Consciousness as a term used by quantum physicists is a more all-encompassing description of an energetic force and includes all sensory perception. So, for example, our brains are aware of the motor traffic around us long before changes in the expected traffic flow are registered in our awareness. Background subconscious processing is going on constantly and our brains only flag to our attention the information which has significance and potentially requires action.

Furthermore, we now understand this broader human consciousness to be a primary force, not a secondary effect. Cast your minds back to the role of the observer in the double-slit experiment: the wave form materialised as a particle only when observed. Objects become perceivable at the point of intersection between our consciousness and the energy wave form we are observing. Think of an arrow in flight. This is an energy system in motion. We cannot 'see' the arrow in motion. We just make mental photographs of its passage through space and link those together to give us a perspective on its journey. All we know for certain is that it arrives at its destination at the point of impact. We could not have been certain that it would get there until it arrived. Our minds take snapshots at the points of intersection between our consciousness and the energy wave of the arrow. If there is no consciousness, there would be only invisible energetic waves. Therefore, consciousness makes things happen, including perception of the universe itself.

The rapid momentum of neuroscience is producing more and more evidence that the brain is designed to interact with the energetic universe. We recognise synapses as the points where neurons interconnect physically (wire together), but they also operate as a bridge to the energetic domain, a configuration of energetic fields associated with the excitation of ion channels in the plasma membranes of the neurons. Technically we would refer to this as a spatial-temporal dimension of neural firing. For example, we know that memory cannot be fully understood without reference to quantum energy. Although there are parts of the brain that specialise in certain aspects of memory function, such as the hippocampus, the 'filing cabinet' of the brain, it cannot be pinned down to one area. Memory works by leaving energetic residue ('quantum footprints') across the brain, which act as energy markers to trigger dormant neural configurations.

We also know that consciousness is not just about the brain: the heart is directly involved in stimulating consciousness via the vagus nerve, the neural connection which runs inside the spinal column. Our bodies have a sophisticated network of somatic markers, which are cell structures that sense the outside world and signal the brain to intervene when certain energy thresholds are hit. And so, the evidence accumulates that our physical dimension is only one part of who we are, and that we have brains and bodies that are also designed to plug into the energetic environment which surrounds us.

There is a field of scientific research known as noetic science. This is concerned with bringing the rigorous methodology of science to the study of the paranormal. It has been spearheaded by Dean Radin and has used well-proven statistical techniques to analyse large amounts of accumulated research data on phenomena such as collective consciousness, clairvoyance, prayer, meditation, parapsychology, and healing at a distance.

In its early years, such research would have been easily dismissed by the wider science community as lacking credibility, but its rigorous approach and the parallel findings in other fields have contributed to an important change of attitude.

Radin's research shows that paranormal phenomena can no longer be dismissed as irrelevant. Of course, there are wild claims that can be easily filtered out, but when we get down to the serious stuff, we can see that certain people are capable of producing results, such as future predictions or healing, which cannot be explained by classical science alone. Crucially, they cannot be dismissed either. The earlier narrowness of mind came from the fact that paranormal findings are not repeatable in the way of classical scientific methodology, but neither are they random. The results show many examples of the chances of paranormal successes being the result of random factors at around one billion to one against. In other words, the results were not random. So, what forces were in play?

States of consciousness

Consciousness is a state which fascinates both neuroscientists and quantum physicists, so let's dig in a little deeper to understand why. In an earlier chapter we referred to consciousness as a personal sense of being. When we explore further, we can find this is not as constant a state as we might like to think. Here we look at different states of consciousness and start to ponder what this means for our newly emerging understanding of human nature.

Consciousness is not a physical state, but we are able to understand various levels of consciousness through measuring brainwaves and their frequencies. Our brains constantly emit energetic waveforms at different frequencies. These tell us a lot about brain states and are typically monitored by an EEG, basically the brain version of the more familiar ECG, which is used for heart rhythms. Brainwaves are transmitted across the brain and reflect various levels of activity. At the lowest frequencies (2 to 3 hertz) we have delta waves, which reflect a state of deep (non-REM) sleep. Next comes theta waves (4 to 8 hertz) where we dream and exercise creative thought. Our standard operating mode during wakefulness occurs within the alpha range (9 to 12 hertz). Then more intensive cognitive and rational processing occurs within the beta range (13 to 38 hertz). Finally, higher thought processes, which include intuition and imagination, take us into the gamma range (38 to 42 hertz). These distinctions are important in helping us to understand what is going on energetically within the brain (see Figure 12).

BRAINWAVES

Hz	Range		
38 - 42	GAMMA	Higher thought-processing	Openness (internal)
13 - 38	BETA	Arousal, engagement, attention	Analytical (external)
8 - 12	ALPHA	Default operation/Consciousness	Normal operational
4 - 8	THETA	Dreaming/ day dreaming	Creative
2 - 3	DELTA	Non- REM Sleep	Rest

Figure 12: Brainwaves

Consciousness is not a state of singularity or a constant. It is a dynamic state which is always evolving, a field of energy perceiving, interpreting, and responding to our worlds – a stream of consciousness. We may feel that self and consciousness are one and the same thing, and this is arguably true in any one moment, but the self of the past is not the self of the future. We are not a fixed entity making our way through life. We are a process.

The more we look into this, the more any notion of constancy of self disappears. I believe this is really important for us to understand as human beings. We are not a constant self. Ultimately, we are not tied to anything we might have been in the past. Each of us is a dynamic force that has the potential to change its course: the challenge is that scientifically we are only just starting to understand how we might do this. It is this potential to be so much more than we are now that is emerging before us, and the secrets are held in the energetic domain.

I want to reinforce this perspective by exploring a little further some examples of what we describe as 'altered states of consciousness'. These show how frail and temporary the state of consciousness can be and how any tendency to attach our own sense of selves to any one snapshot of consciousness is futile.

Out-of-body experiences (OBEs) are well-documented, particularly within the medical community and especially within surgery. They represent a version of out-of-body consciousness. There are many examples of recovering surgery patients reporting a sensation of being out of their bodies during their operation. Typically, this would have felt like floating up at the ceiling and being able to observe all that was going on below them. In many cases, this has been explained away mainly as a hallucinogenic effect of the drugs that were used. Many drugs can have this effect and so the explanation seemed plausible, but the pattern of brainwaves did not match the known effects of the drugs. Rather than being in the expected high range of brainwave frequencies, their brains were in the low range, yet accompanied by very lucid thinking.

The second easy explanation was that the patients must have noticed these things in a preconscious state before the surgery got underway. However, a significant number of patients were able to report things that they could not possibly have observed from their vantage point, such as reporting the brand name shown on the underside of the operating table!

Patients reporting OBEs explained a sensation of total observation, as though they could see everything that was going on in the operating theatre at the same time. It were as if they had freed themselves of their bodies and assumed a totally energetic state. For the moment, no satisfactory explanation has been proven.

Premonitions are a form of predictive consciousness and also fall into this category of not proven according to repeatable results, but neither dismissible as random. There are many documented cases of people predicting the future with amazing levels of accuracy. Abraham Lincoln explained a dream 13 days before his assassination, depicting in detail his demise.

Extrasensory perception (ESP) is an example of connected consciousness. Significant research data has been reported in controllable laboratory conditions across thousands of cases. Many identical twins appear to share such a connection, including the case of the three-day old twin who spontaneously burst into a terror fright, only to be consoled when his brother in a separate cot was discovered to be suffocating and was revived in the nick of time.

Then there is the perennial discussion of mind over matter, a form of kinetic consciousness. This is no longer an 'if' debate, it is a 'how'. There is widespread evidence showing the impact of mental forces on the health of our bodies. By concentrating our focus on a particular part of the body, we can cause the temperature of the blood in that area to rise allowing better flow of nutrients which aid health. In particular, the placebo effect has been well researched. This is where 'phantom medication', such as the use of sugar pills, is used to treat patients. The patients are left in the dark and still believe that they are being treated by authentic drugs. The typical effect of such belief is a measurable benefit in their condition. Also, in laboratory conditions, Tibetan monks demonstrated their meditative powers to alter the molecular structure of water. Plants have been shown to flourish when they are 'sent love' by trained researchers, whereas the progress of the control plant group remained neutral.

Then there is the other side of the spectrum where people lose some of their mental faculties through aging diseases. Dementia is the classic example. Anyone who has witnessed the progression of dementia or Alzheimer's in a loved one will testify to the destructive impact it has on memory. This can impact the ability to carry out many tasks as access to the memory circuits is blocked by the build-up of plaque at the synapses. What is also striking is the effect on the person's consciousness. The person's former personality gradually disintegrates as their access to memory narrows, typically with more recent memories disappearing first and older memories later. Their former selves are lost in the destruction of their consciousness.

Collective consciousness was conceptualised by Carl Jung, the famous Swiss psychologist, in the early 20th century. This postulates the existence of a collective mind that sits outside of all of us, where our thoughts converge into one arena of consciousness. Today there is considerable evidence to support the existence of such a phenomenon, a realm of collective consciousness operating at the top end of the gamma range, around 40 hertz. As explained earlier, this is the range of higher individual thinking where our intuitive and imaginative capabilities reach their peak. It also appears to be the threshold at which the gateway to collective consciousness opens up to us. Quantum physics recognises Schumann resonances, an electromagnetic field surrounding the Earth which appears to impact collective human health. While we are in sync with this field our health improves, we deteriorate when out of sync. There is something invisible going on all around us, and we are only just starting to develop the scientific explanation.

The next piece of intriguing evidence comes from split-brain experiments. This sounds pretty scary but comes about only when patients are already suffering highly debilitating brain conditions. For instance, in cases of uncontrollable and constant brain seizures, neurosurgeons can intervene to separate the two main hemispheres of the brain at the corpus callosum, which is the main neural connection bridge between the hemispheres. This has the effect of reducing the overall level of electrical activity across the brain, hence rendering it more controllable. In effect, such patients to some degree now have two brains. The implications for consciousness are fascinating. What is seen is the emergence of two consciousnesses. The patient presents like two personalities, sometimes collaborating and other times conflicting. While not an ideal outcome for the patients, it was much better than the impossible lives they were experiencing beforehand.

There is also the amazing case of Jill Bolte Taylor, a neuroscientist who suffered a stroke in her 30s. Here was a scientist witnessing the collapse of her brain function from the inside. This is beautifully captured in the YouTube video, "My Stroke of Insight". She talks about her experience of the splitting of consciousness between her two hemispheres, the right side enjoying the energetic release of feeling part of the universe, and the left side trying to hold on tightly to its place in the physical world.

This concept can be extended further with a thought experiment. If we were to wire the brains of two people together, both consciousnesses would merge into one. The former consciousness of each person would disappear, as would their personalities.

The state of hypnotic trance also throws up some interesting insights as a further example of an altered state of consciousness. Hypnotists use techniques which enable their clients to relax into the theta range of brainwave frequencies. This is the range where we experience lighter sleep and sleepy wakefulness, or daydreaming. This is where we are at our most suggestible. Intriguingly, particular types of geometric shapes and images can be very successful in inducing trance. These can work like a kaleidoscope, a dynamic mixing of recurring shapes to create ever new visual patterns. Fractal patterns are recurring shapes which appear across nature and have been occasionally afforded important spiritual significance, referred to as 'the sacred geometry of the universe'. Tangible examples are snowflakes and honeycombs. Where do the designs come from? This takes us back into quantum science at the most fundamental level.

When two chemical elements combine, they must communicate in order to exchange information before they can change their behaviour. Take the example of water. If we put two particles of hydrogen with one of oxygen, they have to connect in order to transform into a molecule. There has to be a transformation process. This chemical reaction comes about when the energetic frequencies of the atomic elements resonate at a particular frequency. When this happens, encoded information is carried in these frequencies which allow this communication to take place. Although still surrounded in mystery, it appears that such energetic fractal patterns are the universe's means of exchanging information, the language of the universe. This may explain why such patterns are effective in trance, as though they enable some deeper connection with nature at the level of the quantum field.

On a more easily relatable level, our own dreams offer us

insights into consciousness. We dream throughout the night in accordance with our circadian rhythm, which acts as the time clock of our bodies. During our early deep sleep phase, we have more thought-based dreams, and in the second phase our dreams become more emotionally intense and creative as our limbic system more directly engages. The key neurotransmitter causing sleep is melatonin, which suppresses the normal reactions of the thinking mind and our connections to our motor neural responses. This is in effect a state of temporary partial paralysis. I guess we have all experienced dreams where we are trying to do something, but we can't see ourselves physically doing it, such as running away from a threat but our legs won't move us fast enough. We also have probably all dreamed of being weightless, flying or floating around various scenes without being fettered by the physical limitations of our bodies.

Just like the trance and dream states discussed above, meditation takes us into an altered state of consciousness. It suppresses the standard operating mode of the brain, which occurs in the alpha range of brain frequencies, as well as the rational and analytical processing of the beta range. We drop into the theta range, which allows a deeper connection with our intuitive selves uncluttered by our normally busier minds. As meditative practice is mastered, brainwaves become more prominent at the gamma range, the level of highest thought. This is consistent with people experiencing deep and powerful insights which would have been out of reach of our normal cognitive processing. As meditative practice strengthens further, we appear to enter the realm of collective consciousness. Spiritual gurus who have mastered the techniques of such practice appear to enter a new state of being and show very little appetite for returning to the confines of the material world.

Meditation is, of course, a subjective experience, but it is not out of the reach of scientific study. Apart from consistently proven health benefits, especially in relation to cardiac and respiratory health, we can also demonstrate significant positive impact on emotional wellbeing. We are starting to understand how this works. Melatonin has a significant role in inducing meditative states. We also know that the pineal gland in the brain, when stimulated by passing cerebrospinal fluid, acts as the gateway to wider energetic connection and potential experiences of collective consciousness. The pineal gland becomes our own Wi-Fi system, heightening our internal antennae to connect with external energy. It also acts like a transducer, converting these energetic signals into internal imagery, just like the television picking up broadcasts and converting them into the images on our screens.

I hope this whistle-stop tour will help you to start letting go of any notion that we are just one state of consciousness. From a scientific perspective, we know there is no constant self, and consciousness itself, while an ever-present force, cannot be reduced to one state. So, we come back to the question: who are we at source? Science does not have the ultimate answer, only insights which we can personally piece together to make up our own minds, which is where we return to the subject of belief.

Spirituality and belief

As human beings we have pondered the unknown for many millennia. Cave art depicting gods and spiritual symbols has been found dating back to 50,000 years ago. This fascination continued with the Egyptians and Greeks and has been an ever-present aspect of the human psyche ever since. Today we see religions and wider spiritual beliefs practiced throughout the world. As we try to cast a scientific view over this, some interesting questions emerge.

Are mystics and spiritual gurus really able to see and experience something that the rest of us cannot? We could dismiss them as delusional, and while that may be true of some, I doubt whether it is true of them all. Science is showing us that meditative practice can elevate our ability to connect with higher consciousness, to a realm where we can see or sense a level of understanding that could be described as sacred in the richness of its texture. We can match the reported subjective experiences of meditating monks with the brain areas that are activated at the time, namely, the parietal lobe (the top part of the cortex) and the prefrontal cortex. We can see the more intuitive right hemisphere of the brain light up as the more analytical left hemisphere calms down and brainwave frequencies move towards the gamma range of higher processing. We now understand that the more inward-looking right brain holds the key to these higher gateways, while the outward-looking logic of the left brain provides a barrier rooted in the physical world. Science is starting to make inroads into the realm of these unknowns.

The key to meditative practice appears to be learning to go beyond the self, to escape the story we have been telling ourselves throughout our lives as to who we are. In an uncertain world, we cannot cope with ambiguity about our own identity, and consequently we cling on to a belief system that we hope will protect us through life. The price of this self-conceived fantasy is that we blind our eyes to the wider potential that surrounds us. The act of knowing is, by its nature, a filtering process that blocks out what we consider to be meaningless information. But if our interpretation of what is meaningful is too narrow, we limit our opportunity to grow. Our perception of the universe is entirely created within our heads. We are not perceiving an independent external reality; we are representing it within our own internal sensory and cognitive systems. We co-create its shape and form through our own consciousness.

So, once again back to the hard question: who are we at

source? I claim no expertise on this whatsoever nor that my view has any more significance than anyone else's. Yet, I have my own need to try to make sense of this. I used to buy into the idea that each of us is the 'observer' of our own experiences. We are not any of our thoughts, feelings, instincts, or memories: we are the whole, the essence that watches them pass by in the passage of our lives. More recently my view has evolved further. With the understanding I have today, I believe that we are 'the experience itself'. We are the unique flow of subjective experiences that constitutes the journey of our life. Nobody else has lived this life and nobody else can share exactly the same perspective on what life is. We are unique observers, but we are more than this. We are immersed in life and are the experience itself, each of us occupying our own very special place in the universe.

If the experience is our essence, what is the soul? The term 'soul' infers a form of being that exists before human life and continues after it. Once again, there is no scientific answer to this question. The idea of a soul emerged across early religions as a means of making sense of the apparent void we call death. It became a belief that offered us the comfort of a continuous self, bridging the boundary between life and the unknown. By saying this, I am not dismissing its value or denying its authenticity in any way. In fact, I think the quantum parallels are striking.

Physics tells us that energy cannot be created or destroyed, only transformed or channelled. So, at the point of conception, where did the energy of our life force come from? And where does it go when we die? Quantum theory maintains that quantum energy is the source of all things, so it would follow that our life force comes from this quantum realm of infinite potential, from where it materialises as human form at the point of conception, and to which it returns when our life force is no longer anchored in the physical world at the point of death. David Bohm, one of the most influential quantum physicists of the 20th century, referred to the quantum realm as a place of oneness where everything is connected: "The universal wholeness in flowing motion." Hardly the turgid language that many may expect of a physicist! Is this a reference to heaven?

In most religions, there are references to the spirit becoming manifest in human form, and a belief that we come from a divine realm and return there when our life journey is over. Are the 'divine realm' and 'universal wholeness' not the same place? When humans conceptualised the notion of God or heaven, were they simply reflecting a deep collective human need, or is there a place that we can all access at the deepest subconscious level where we have freed ourselves from the limitations of self and physicality to know that these things are true? And to take the argument further, when we die, if our energetic essence returns to the universal realm, do we take with us our experience of life in the form of encoded energetic information? Is this the means by which every life contributes to the essence and momentum of the universe?

This convergence of thinking is especially evident within metaphysics, where philosophy meets science to explore such big questions. Quantum physicists regard the speed of light as the 'point of separation' between the perceivable and the unknown, between the observable physical world and the invisible energy of the universe. Quantum reactions occurring at the subatomic level happen significantly faster than the speed of light and are therefore outside of our range of perception. So, quantum physicists work with probability theories and formulae to predict the dynamics of the unknown. Is probability the bridge between science and faith?

There is another crossover for me in looking at miracles and synchronicity. Many religions depict miracles as a representation of the power of belief in the spiritual realm. In the early days, we did not have noetic science to analyse these, so it is likely that they would have been embellished by storytelling as the main means of informing the next generation. But miracles continue to happen every day all around us. The dramatic exceptions grab the headlines, such as the hospital patients who have survived deathly illnesses against all odds and contrary to any reasonable medical explanation. Isn't every birth a miracle? Isn't the process by which two cells combine to become a fully functioning human being a marvel? It seems to me that we lose our sense of the wondrous through familiarity and the miraculous nature of life itself gets taken for granted.

When we look at the scientific take on this, miracles might be explained as the consequence of imperceptible subconscious interactions between ourselves and the quantum realm. The universe is an infinite field of energetic forces that intersect without limitations of space or time. When mystics reach higher levels of trance, they consistently explain experiencing the universe as being timeless and sensing all things that ever were and will be. Einstein would have concurred with such a hypothesis. Religion has depicted many prophets with miraculous powers. Were they able to reach a perceptual vantage point in higher consciousness that allowed them to look out from a higher plane and which enabled them to see the possibilities that can potentially unfold ahead of us? Did the most special of them learn the power of channelling this potential through belief to make the unbelievable happen?

The state of belief is a deeply personal phenomenon, a necessity for making sense of our lives, with the potential to make miracles happen or to lock us into a prison of stagnation. Whatever we believe, the universe keeps moving forward. We as human beings represent a very special place in its journey. The human brain is the most sophisticated system we know of anywhere in the universe. We have the capacity for more connections in the brain than there are known atoms in the universe. We have the ability to create new fields of meaning through consciousness, allowing the universe to discover a new dimension of learning. We are the only species with the ability to consciously observe the universe and its wonders and to provide intuitive, experiential feedback. We have arrived at this place through both science and belief. Are we not miracles ourselves?

CHAPTER 9.
TRANSFORMING SOCIETY

We have built a world without trust, yet the desire to trust is a fundamental feature of all humans. We are designed to bond together in communities, but many of our problems arise from not having a global vision, which would in turn give us a global purpose. The globalisation of our economies means that we live close enough to be seen to pose a threat to each other. In the territorial world of distrust, neighbours are seen as potential invaders, and everything we own within our territories can be taken away from us by those invaders, so we have to fight to hold on to what we have. The demise of others is seen simply as proof that it could be us next.

I am not adopting a political standpoint here but am simply pointing out that our current global structure (or lack of it) leaves a void that creates conditions for the most destructive aspects of human nature. Think instead of a world where we feel safe, where we are allowed to follow the natural expression of our talents, where equality of opportunity is a given, and we share mutual respect for nature and across communities. I know this sounds like a (very good) John Lennon song, but I do think it is achievable. Of course, it could take several generations, but in a world already tasked with transforming its attitude to our climate and environment, I think we should also look again at how we handle ourselves.

The notion of some form of multitiered global government is a serious one. Maybe a significantly improved version of the

United Nations, but where agreed collective values are the driving force of its stewardship, and a relentless pursuit of transparency and trust is its purpose. It needs to be built on knowledge that has learned from the mistakes of history. Setting up the machinery of such a government is well within our capabilities if the will is there, and I suspect this is where the biggest barrier lies: asking those who have power now to give it up for a greater good is indeed a major obstacle to overcome. Power corrupts and therefore we should be very careful about whom we give power to and how they are held accountable. In my experience, people across the globe ultimately want the same things; it is the territorial leadership which divides us, not human need. Ultimately their power depends on our collective consent. Any organisation has to provide purpose and connection to those it wants to lead. There are many recent examples of central governments becoming meaningless to their people, and people who feel unrepresented will look to another source of purpose to align their loyalties and identity.

Likewise, our economies are by their very nature transactional, designed for the practical purposes of trade. While there is value in this, it is soulless. What price for relationships, trust, collaboration, social value, and community loyalty? There are ways that economies can be transformed to align with the greater goals of human progress. In my opinion, greater value needs to be assigned to jobs of social nature, jobs which can support us to take our talents to new levels and to look after each other. There is a huge imbalance created by those who make fortunes through purely individualistic pursuits. Some would argue that such individual motivation is vital to maximising economic value. This is probably true in the current transactional economy. However, I believe those who pursue their own success are themselves ultimately seeking validation from others, just like the rest of us. The problem is that in today's society we are taught to honour economic and material success at the expense of emotional wellbeing and social connections. We need to rethink this.

In particular, there is a disastrous tension between economic consumerism and environmental care. The success of the prevalent current business model is built entirely on growth. Industrialised businesses which do not grow do not survive. This means that we are constantly taking resources from our planet to feed this insatiable consumer monster. Unlike nature, it does not stop eating when it is no longer hungry. It keeps coming back for more. Consider the incredible amount of money spent globally on fanning the flames of consumer demand. We see it every day when we switch on our televisions, smart phones, or computers. Regrettably those same flames are destroying our planet, eradicating species, and damaging natural habitats and biodiversity.

I am struck by the equally soulless nature of our legal systems, but I do not blame the legal practitioners (well, maybe some of them!). The law is overburdened by the need to try to legislate for every possible aspect of human behaviour, an impossible task. Why? Because we have by default tried to fill the void of distrust with bureaucracy. Justice is an incredibly important aspect of community life, but it is a sense, an emotional value which sits deep within our brains. It can be supported by legal frameworks but not replaced by the detail of rational rules. Maybe we should be looking again at the modern equivalent of community elders, tasked with the role of dispensing justice and wisdom, and accountable to the people they represent. This would not work today in a swamp of social and political distrust, but perhaps it could play a role in the future?

Within the business world, there is an equal need for transformation. Corporate structures and practices are rarely fit for purposes of longer-term social and environmental sustainability. Faceless shareholders whose only interest is in financial gain do not belong in the future vision emerging here. I have no problem with people making money; it is a vital aspect of economic sustainability. My problem comes when such gain is achieved at the expense of others. There are ample opportunities for collaborative economies and non-competitive business practice if only we have the collective will to grab them. Having said that, I do detect a shift towards more ethical investment and am pleased to have met and engaged with a number of wealthy people who are keen to invest in a better world for all.

And, of course, education sits at the heart of this transformational challenge. We need to move away from the current focus on developing children as economic units, cannon fodder to be fired into the machinery of transactional businesses and economies. Intelligence and social value come in many forms, not just the rational process of passing examinations. Rather than being prescriptive about what makes a child educationally successful, we should be exploring the talent their unique personalities can bring to their world. Children who do not fit the prescribed criteria should not be branded failures and left on the scrapheap of rejection and stigma. Every human brain has infinite potential and will only fail because they have been let down by the society that is supposed to serve them. The education system should be a process of discovering and developing young talent and matching it with the right social needs, including industry, art, science, exploration, health, and social support. No talent should be shunned or wasted.

The challenge of creating a new future comes at a time of amazing technological advances. Transformation is happening now in our daily lives as technology changes what we do and how we do it. While the speed of change may be unique in human history, the nature of the change is not. Once again, we are faced with questions about the role of humans in societies where so much more can be done by machines. In the past, machines have taken over many aspects of manual labour. Now they are moving up the value chain. All rational processing tasks are up for grabs. Anything which can be reduced to binary definition is within scope, which is a massive aspect of our current lives. Like all technologies, they can be both used and abused. The hands that navigate the progression of such capability need to be wise and compassionate.

This is particularly true of social media. The level of cyber bullying is terrifying. Having children and young adults exposed to such abuse has already wreaked havoc and pain for so many. The accelerated use of intelligent customisation to user preferences makes complete business sense in terms of marketing success, but what of the dangers of reinforcing prejudices? If the material we are offered merely strengthens the narrowness of our understanding, what hope is there for wider empathy and mutual tolerance? Here is another example where connection is vital. If the social media giants have global ambitions then global responsibility has to come with it, including enlightened collaboration with the governments of the day.

In parallel, sophisticated social media and artificial intelligence could present a threat to our individuality. Technology is constantly observing and predicting our behaviour. We are tempted to follow the behaviour patterns already analysed by the huge computing power that we have created. Can these suggestions and prompts become irresistible forces of influence and eventually control? Will we become more locked than ever in patterns that imprison us? Will we become the rats in the experimental maze?

On the other hand, technology could play a transformational role in democracy and government. Of course, it already is, as no government can afford to ignore the power of social media. The current problem is that there is no global government force sufficiently harmonised to channel the benefits this could bring. Consider democracy: we already have the power to conduct instantaneous mass surveys of people sentiment across the globe. Sophisticated reports can be turned around in hours with amazing granularity and insights. Why then should we rely on the slow-moving democratic processes that are currently in place? Proper harnessing of this consultative value could and should revolutionise political representation and bring about a more immediate, responsive, transparent, and accountable representation model.

History tells us there is no point in resisting this momentum, but there is every reason to channel it. While our rational capacities may be delegated more and more to machines, this still remains only a small part of who we are. Maybe the manmade advances of artificial intelligence and robotic machinery will prove to be a further catalyst to our own accelerated evolution to a next stage of human potential? We still have the infinitely rich mix of intuition, emotion, and instincts to define us, capabilities which remain well beyond the grasp of modern technology. We may have robots that look like they can be emotionally intelligent, but they are not; at least not like we are. We can feel and sense each other, experiencing genuine empathy and bonding – a vastly richer experience than the rational and pattern spotting formulae of robots.

Rather than agonise over the parts of us that can be replaced, surely it is better to concentrate on what we can move on to, exploring the realm of untapped human potential that awaits our engagement. Like the universe, we need to keep moving forward, opening up new areas of learning and finding new connections to enable our progress. Higher collective consciousness appears to hold the key to the next level of our evolution. Humanity will continue to unfold in the belief that we are part of something bigger and that our potential is limited only by our perception.

Our focus on materialism has created a world of separation. We have been encouraged to see ourselves as individuals, to view relationships with scepticism, and to see idealism as a weakness and inspiration as a fantasy. We have seen the planet as our property to be used or discarded as we see fit. The quantum way acknowledges the deep connection between us and with nature and recognises that real sustainable progress can only be achieved when all these needs are respected and nurtured. When we damage nature, we damage ourselves. The two are ultimately inseparable. Being in tune with nature has been seen for too long as a luxury and somebody else's problem. We need to wake up and recognise this as a prerequisite to human sustainability, a need which should be acknowledged from the heart of our collective consciousness.

Quantum thinking is holistic and uses every aspect of our combined capabilities and talents. We have to learn again to trust such wider forms of inner intelligence. Throughout this book I have pointed out the dangers of limiting ourselves to rational thinking alone, and instead have given equal credence to the intuitive and the emotional. Yet, it is the rational preoccupation of science that has enabled us to view the universe at its most fundamental quantum level. The tools we created to understand the material world have actually taken us into the quantum realm. This would not have been predicted or intended 100 years ago. The path of human enquiry is neither linear nor predictable; it unfolds and opens up many new roads to knowledge and wisdom. Our task is to stay open to all of these wonders.

To adapt a phrase typically used in a different context, 'the universe works in mysterious ways'. It brings together its infinite mix of intelligent resources to keep moving forward, to overcome obstacles, and, where needed, to find a better way.

Now is a good time to remind ourselves that we, as human beings, have the capability to do the same.

CHAPTER 10. WHAT CAN I DO NOW?

You may be asking yourself: what can I do now?

Much of the content of this has been conceptual and reflective, so the challenge is how to ground these principles in tangible actions. I have to confess a personal prejudice which is that I am wary of solution-oriented models that suggest that all you have to do is follow their rules and success will follow. In this book, I hope I have provided you with a blend of insights that can support you on your own learning journey, but there is nothing linear about this process. It is up to you to find the blend that suits you most. Every journey is personal.

On the other hand, I do not want to leave you hanging. So, I have suggested in this final chapter possible areas where you may want to begin to develop your own learning. These suggestions are split into two main categories: personal and organisational.

Personal actions

If you are serious about taking on board some of this learning, you must spend time reflecting on yourself, your habits, your reactions, your memories, and the influences in your life. You can choose to do this through a disciplined contemplative or meditative process. Or you can simply get into the habit of creating short windows of time for thinking about your experiences. Remember, the longer you leave the reflection, the more likely you are to post-rationalise the event and to slip back into your normal comfort zone. The closer the reflection is to the event, the more honest you are likely to be. Anger, for instance, will allow you no reflective space in the moment, but the more skilful you become at removing yourself from the immediate emotional hijack, the closer you will be to the truth of your emotions. Why did I react the way I did? What was the trigger to my fear? What was I afraid of? Where in my memory did I learn to be afraid of this type of trigger? What was the chain reaction in my body and what did I see in my mind?

All of these questions are critical to raising your awareness of your personal dynamics: what is going on inside of me and outside of me? It is always about your interaction with your environment and the way we have learned to process personal meaning.

Share your reflections with trusted others. Your friends and colleagues are facing many similar challenges. You can support each other to bring sense to situations and to open mutual doors to heightened awareness.

Learn to see the world through the eyes of others. Get into their headspace in order to understand them. Everything about them has been learned. Digging your heels into your own perspective will confine you to narrowness and prejudice.

Recognise your own habits and walk away from those which no longer serve you. Create room for new habits that will help you on your journey. Stop waiting for perfection. Get on with it now with baby steps. It may feel forced and clumsy in the early days, but if you are authentic, you will find a way. The journey is your teacher. Trust yourself and stay open to the possibilities which will unfold.

Create your own strategy for personal growth. Work at your vision and your purpose. You need a compelling sense of direction. You can work out some of the details on the way. Think about when you are in play, when you are learning or practising, or when regenerating.

These suggestions are not rocket science. If you have the will to move forward, keep that energy alive; show it the sunlight and let it take you to places you may not have predicted.

Organisational actions

If your interest is organisational, here are some suggestions you might choose to follow.

The timing for change is perfect. Covid-19 restrictions have given many people the opportunity to stop and reflect on the type of life they want to return to. The pandemic has caused the ancient pillars of traditional organisations to crumble along with their insistence on place and control. We have all witnessed new possibilities offered through remote technology and the associated opportunity for personal freedom. Suddenly, we found the rules of engagement changing before our eyes on a global scale, and we were all in it together. Organisations should take the opportunity to think through very carefully how they define the phoenix that is to rise from the ashes.

There are a few frameworks and tools scattered throughout this book which are designed to support the process of transformational change. In particular, getting to grips with the behavioural styles covered in Chapter 3 will prove a very productive investment. This will support people to understand themselves better, as well as their interaction with their colleagues. Most importantly, it will heighten awareness of brain and energy dynamics, providing a common learning language that can be shared across the organisation.

If you take on the challenge of becoming a quantum organisation, you will need to create a programme which has the capacity and energy to transform, and this can only be achieved by mass engagement of all people in the business. In my experience, change programmes rarely pass this test. Their programmes are normally unduly focused on the few, and the rest remain sceptical bystanders. Moreover, this engagement has to become a daily experience, not just an occasional workshop.

At the outset, the strategic design principles and approach have to be fully debated and agreed. Scalability will be an important consideration for larger businesses, but the principles explored herein are relevant to any size of organisation. Young companies have the advantage of going for the quantum way from the outset whereas most corporate businesses will face the bigger challenge of 'unlearning' the way they have done things in the past. Such a transition has to be thought through very carefully. Larger organisations will need to recognise the wider infrastructure and support requirements and design their programmes accordingly. Early pilots across different parts of the organisation may be a sensible starting point before going on to further phases of wider rollout. This programme is not a quick fix. It is a commitment to a better way of working that plays to people's natural strengths and needs.

I believe there are then three prerequisites to the success of such a programme (see Figure 13).

CULTURAL TRANSFORMATION PROGRAMME

Figure 13: Cultural Transformation Programme

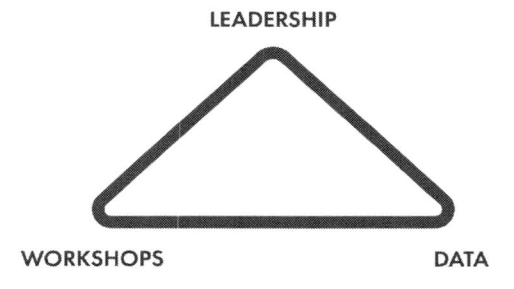

Leadership

We dealt with the features of quantum leadership in Chapter 6. It is vital that leaders across the organisation totally embrace these ideals. They need to have an intuitive connection with its purpose and believe in its potential. They must be committed to working with the insights on a daily basis and finding their own ways to personalise their learning. They will grow throughout the journey and become invaluable role models to those who follow.

Workshops

Regular workshops will serve to deepen the learning and act as ongoing catalysts for motivation and alignment. Although offering initial explanation of the core insights, over time they should become self-generating and create a safe, facilitated space for open exploration and reinforcement. Typical subject areas covered in such workshops reflect the subject matter of this book and would include science insights, the quantum organisation, behavioural styles, leadership, understanding the human experience, building trust and engagement, handling feedback, building resilience, high-performing teams, self-discovery, persona, change, and the science of belief.

Data

Such a transformation programme needs to constantly have its finger on the pulse of its people's emotions, a daily dialogue of feedback and sharing. There is some amazing employee surveying technology available today which can make this an immediate reality. All training faces the challenge of translating the sessions of the classroom into everyday working practice. In this type of programme, such data holds the key. I have found through my own experience that such information sharing enables wide people ownership of the learning agenda. As people discover what others are thinking and feeling, what they are excited by and where they are uncertain, there emerges a connective force which propels the acceleration of cultural development.

In the early days of the programme, people will need support in understanding what will at first appear to be unfamiliar concepts. But that will quickly change as they start to recognise many of the insights as natural states that we experience every day. Their appetite will grow. The principles of connection, transparency, communities with a purpose, and personal growth will inspire a sense of total ownership.

People will need to see authentic leadership letting go of the reins of hierarchical control and learning instead to trust, inspire and enable their people. The organisation will become an aligned, energetic force that will surprise us all as it unlocks its own potential. This is who we are meant to be.

Signing off

My aim in this book has been to challenge your thinking and to do so with a scientific perspective. We should not be afraid to address the unfamiliar when there is serious evidence that there is something significant to be gained by looking it directly in the face. You will not be alone in this journey. Others around you will also find the challenge daunting. Letting go of ingrained thinking does not happen overnight, so please be content with baby steps in the early stages. Bigger energetic shifts await you, but you will need to find your own way to this deeper level of meaning and insight. Don't rely just on your brain to navigate the path; your heart and gut will help you shine a light along the way. We are so much more than we have allowed ourselves to believe, and a new horizon awaits. I wish you huge success in your unique quests.

BIBLIOGRAPHY

Chapter 2
There are several video clips available on YouTube which demonstrate this experiment including "Double Slit Experiment explained! by Jim Al-Khalili" from February 2013, YouTube.
Lanza, Robert. *The Grand Biocentric Design: How Life Creates*. Dallas: BenBella Books, 2020.
Lipton, Bruce. *The Biology of Belief: Unleashing the Power of Consciousness, Matter & Miracles*. London: Hay House UK, 2015.
Chopra, Deepak. *Synchronicity: Harnessing the Infinite Power of Coincidence to Create Miracles*. London: Rider, 2005.

Chapter 3
Damasio, Antonio. *Descartes' Error: Emotion, Reason and the Human Brain*. London: Vintage, 2006.

Chapter 4
Eagleman, David. *Brain: The Story of You*. Edinburgh: Canongate, 2016.
Harris, Annaka. *Consciousness: A Brief Guide to the Fundamental Mystery of the Mind*. HarperAudio, 2019.
HeartMath Institute, accessed March 3, 2021. https://www.heartmath.org.

Chapter 5
Lipton, Bruce. *The Honeymoon Effect: The Science of Creating Heaven*. London: Hay House UK, 2014.

Ramachandran, V.S. *The Tell-Tale Brain: Unlocking the Mystery of Human Nature*. London: Windmill Books, 2012.
Van der Kolk, Bessel. *The Body Keeps the Score: Mind, Brain and Body in the Transformation of Trauma*. London: Penguin, 2015.
Doidge, Norman. *The Brain That Changes Itself: Stories of Personal Triumph from the Frontiers of Brain Science*. London: Penguin, 2008.
Seth, Anil. *30-Second Brain: The 50 Most Mind-Blowing Ideas in Neuroscience, Each Explained in Half a Minute*. London: Icon Books Ltd, 2018.

Chapter 6
Zohar, Danah. *The Quantum Leader: A Revolution in Business Thinking and Practice*. Buffalo: Prometheus Books, 2016.
Rippon, Gina. *The Gendered Brain: The New Neuroscience that Shatters the Myth of the Female Brain*. London: Bodley Head, 2019.
Dispenza, Joe. *Becoming Superhuman*. London: Hay House UK, 2019.
Shambler, Gemma, and Dyal, Jackie. "The Quantum Way with Gemma Shambler and Jackie Dyal." *Happiness and Humans*, February 26, 2021. https://open.spotify.com/show/7h6wCWnjsgOIZSR6h0DKbq.

Chapter 7
"Drivers of Happiness Data." The Happiness Index, accessed March 19, 2021. www.thehappinessindex.com.
"Global Employee Happiness Data", The Happiness Index, accessed March 19, 2021. www.thehappinessindex.com.
"Change Curve." The Happiness Index, accessed March 19, 2021. www.thehappinessindex.com.
Kahn, William. "When you inadvertently invent Employee Engagement with William Kahn." *Happiness

and Humans, March 5, 2021. https://open.spotify.com/episode/4ByT5DjtYFh3vVaX0IjFBR.

Kahn, William. "Psychological conditions of personal engagement and disengagement at work." *Academy of Management Journal*, 33, 4 (November 30, 2017): https://journals.aom.org/doi/10.5465/256287.

Phelan, Matthew. *Freedom To Be Happy: The Business Case for Happiness*. London: Happiness and Humans Publishing, 2020.

"Fatigue or depletion." David Lapin blog post, accessed March 19, 2021. https://lapininternational.com/leadership-blog/r-r/.

Chapter 8

Caplor Horizons, accessed March 3, 2021. https://caplorhorizons.org.

Paper Boat, accessed March 3, 2021. https://paperboatcharity.org.uk.

Institute of Noetic Science, Dean Radin, accessed March 3, 2021. https://noetic.org.

RECOMMENDED READING

- Judy Cannato's *Radical Amazement: Contemplative Lessons from Black Holes, Supernovas, and Other Wonders of the Universe* published by Ava Maria Press, 2006
- Deepak Chopra's *Synchronicity: Harnessing the Infinite Power of Coincidence to Create Miracles* published by Rider, 2005
- Antonio Damasio's *Descartes' Error: Emotion, Reason and the Human Brain* published by Vintage, 2006
- Joe Dispenza's *Becoming Superhuman* published by Hay House UK, 2019
- Norman Doidge's *The Brain That Changes Itself: Stories of Personal Triumph from the Frontiers of Brain Science* published by Penguin, 2008
- David Eagleman's *Brain: The Story of You* published by Canongate, 2016
- Annaka Harris' *Consciousness: A Brief Guide to the Fundamental Mystery of the Mind* published by HarperAudio, 2019
- Joseph Jaworski's *Source: The Inner Path of Knowledge Creation* published by Berrett-Koehler Publishers, 2012
- Daniel Kahneman's *Thinking, Fast and Slow* published by Penguin, 2012
- Robert Lanza's *The Grand Biocentric Design: How Life Creates* published by BenBella Books, 2020
- Joseph Ledoux's *The Emotional Brain: The Mysterious Underpinnings of Emotional Life* published by W&N,

1999
- Guy Leschziner's *The Nocturnal Brain: Nightmares, Neuroscience, and the Secret World of Sleep* published by Simon & Schuster, 2019
- Bruce Lipton's *The Biology of Belief: Unleashing the Power of Consciousness, Matter & Miracles* published by Hay House UK, 2015
- Matthew Lieberman's *Social: Why Our Brains Are Wired to Connect* published by Oxford University Press, 2013
- Bruce Lipton's *The Honeymoon Effect: The Science of Creating Heaven* published by Hay House UK, 2014
- Candace Pert's *Molecules of Emotion: Why You Feel the Way You Feel* published by Simon & Schuster, 1999
- Matthew Phelan's *Freedom To Be Happy: The Business Case for Happiness* published by Happiness and Humans Publishing, 2020
- Dean Radin's *The Noetic Universe* published by Corgi, 2009
- V.S. Ramachandran's *The Tell-Tale Brain: Unlocking the Mystery of Human Nature* published by Windmill Books, 2012
- Gina Rippon's *The Gendered Brain: The New Neuroscience that Shatters the Myth of the Female Brain* published by Bodley Head, 2019
- Daniel J. Siegel's *Mindsight: The New Science of Personal Transformation* published by Bantam, 2010
- Anil Seth's *30-Second Brain: The 50 Most Mind-Blowing Ideas in Neuroscience, Each Explained in Half a Minute* published by Icon Books Ltd, 2018
- Bessel Van der Kolk's *The Body Keeps the Score: Mind, Brain and Body in the Transformation of Trauma* published by Penguin, 2015
- Danah Zohar's *The Quantum Leader: A Revolution in*

Business Thinking and Practice published by Prometheus Books, 2016

ACKNOWLEDGEMENTS

My thanks go to The Happiness Index team who have supported me throughout this process, including Matt, Susan, Joe, Caroline, Jackie, Gemma, and Francesca.

Also, a nod to my buddies from the 'Celtic Clan' for their wider support and long-lasting friendship – Diana, John, Jill, Tania, and Wendy.

And here's to the people of Caplor Horizons and Paper Boat who do such amazing work supporting disadvantaged people across the world, especially Ian, Lorna, Rosie, and Kemal.

ABOUT THE AUTHOR

Clive Hyland

Clive Hyland is able to call on more than 30 years of business and leadership experience. Having originally studied sociology and psychology, his interest over the last 16 years has switched to neuroscience and, most recently, quantum physics, both of which he sees as opening up a totally new understanding of human nature and its role in the unfolding universe. The Quantum Way is his third book (coming after Connect and The Neuro Edge) and is a significant milestone in his continuing journey of exploration. Clive is convinced that traditional organisations built on hierarchical control have to change radically in order to create the working environments needed to truly get the best out of people. This challenges all of us to look again at who we are and how we interact with those around us.

Printed in Great Britain
by Amazon